Environmental Footprints and Eco-design of Products and Processes

Series editor

Subramanian Senthilkannan Muthu, SGS Hong Kong Limited,
Hong Kong, Hong Kong SAR

More information about this series at http://www.springer.com/series/13340

Subramanian Senthilkannan Muthu
Monica Mahesh Savalani
Editors

Handbook of Sustainability in Additive Manufacturing

Volume 1

Springer

Editors
Subramanian Senthilkannan Muthu
Environmental Services Manager-Asia
SGS Hong Kong Limited
Hong Kong
Hong Kong SAR

Monica Mahesh Savalani
Department of Industrial and Systems
 Engineering
The Hong Kong Polytechnic University
Hong Kong
Hong Kong SAR

ISSN 2345-7651　　　　　　　ISSN 2345-766X　(electronic)
Environmental Footprints and Eco-design of Products and Processes
ISBN 978-981-10-0547-3　　　ISBN 978-981-10-0549-7　(eBook)
DOI 10.1007/978-981-10-0549-7

Library of Congress Control Number: 2016930176

Printed on acid-free paper

This Springer imprint is published by SpringerNature
The registered company is Springer Science+Business Media Singapore Pte Ltd.

Contents

Introduction

Subramanian Senthilkannan Muthu and Monica Mahesh Savalani

1 What is Additive Manufacturing?

The term '*additive manufacturing*' is the official industry standard term (ASTM F2792) for all applications of Rapid Prototyping (RP) technology. It is *defined* as the process of joining materials to make objects from 3D model data, usually layer upon layer (as shown in Fig. 1), as opposed to subtractive *manufacturing* methodologies. This group of technologies has also commonly been known as 3D printing, Additive Techniques, Layer Manufacturing, and Freeform fabrication.

These technologies are predominantly used in high value-added industries and applications including those involved with aerospace, automotive, and biomedical products which require highly complex and customized designs at low volumes which are currently not effectively met by conventional techniques. In recent years, AM has experienced unprecedented development. Improvements and developments in terms of speed, accuracy, material properties, machine reliability, and, most important, the development of low-cost machines has widened the user base for this technology. Direct fabrication using AM technology is only just starting to be used and we can expect a rapid increase in output as we learn how to employ this technology properly as new markets develop. Currently, just one in a thousand products is fabricated using AM technologies and this technology is gearing towards fabricating a much closer link to the final product. Global manufacturing was worth $10.5 trillion in 2011 and is predicted to be worth $15.9 trillion in 2025. Of this, the additive manufacturing economy was worth $1.7 billion in 2011

S.S. Muthu (✉)
Environmental Services Manager-Asia,
SGS Hong Kong Limited, Hong Kong, Hong Kong
e-mail: drsskannanmuthu@gmail.com

M.M. Savalani
Department of Industrial and Systems Engineering,
The Hong Kong Polytechnic University, Hong Kong, Hong Kong
e-mail: monica.mahesh.savalani@polyu.edu.hk

© Springer Science+Business Media Singapore 2016
S.S. Muthu and M.M. Savalani (eds.), *Handbook of Sustainability in Additive Manufacturing*, Environmental Footprints and Eco-design of Products and Processes, DOI 10.1007/978-981-10-0549-7_1

Fig. 1 Concept of additive manufacturing of a part. **a** 3D CAD geometry. **b** Layered construction

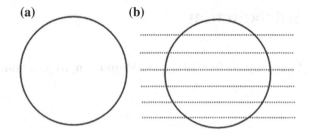

and is estimated to be worth $10+ billion by 2025 (Wohlers Report 2013). This can be seen in the growth of AM techniques and its processing as shown in Fig. 2.

The birth of this technology coincided with the development of Stereolithography (SLA) which was demonstrated in 1987. Today, there are over 30 different processes but the basic principle of this technology remains. The basic principle and steps have common stages. The typical process chain common to most AM processes is illustrated in Fig. 3.

The AM process chain begins by defining a geometry. This can be defined by an object, direct design, reverse engineered data, or mathematical data, independent of the RP system. This information is formulated into a 3D CAD model. To proceed, the CAD file is converted into a standard tessellation language (.STL) file. Since 1990, all major CAD/CAM vendors have developed and integrated the CAD-.STL interface into their system (Chua and Leong 2003a). The .STL files represent a three-dimensional surface as an assembly of planar triangles in either ASCII or binary format. Each triangle is considered as a facet, associated with co-ordinates with vertices and the direction of outward normal with X, Y, and Z to indicate which side of the facet is an object. This tessellated model is used as an input for slicing. Before slicing, it is important that the optimum build orientation

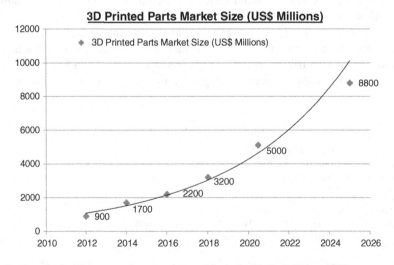

Fig. 2 Growth of additive manufacturing (adapted from the Wohler's Report 2013)

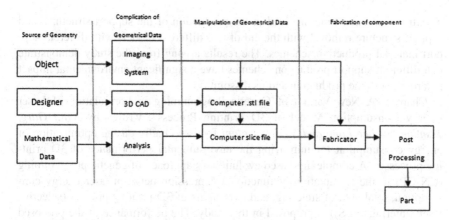

Fig. 3 Main process stages common to most additive manufacturing systems (*Source* Authors)

is determined because the orientation affects the time needed to build the part, its material properties, surface quality, and the need for support structures (Gibson and Dongping 1997). Once, this has been determined, the model can be sliced by intersecting the model with planes parallel to the platform in order to obtain the contours. Slicing can be direct or adaptive, depending on the technique (Gibson and Dongping 1997). To achieve the most accurate parts it is most sensible to use the smallest possible layer thickness. However, this would require more slices, resulting in longer data processing time, larger data files, and a longer build time. Once the parts are sliced, the scan paths are generated depending on the process. Finally, the parts are built.

In comparison to subtractive processing, AM processes are quite autonomous and require minimal human intervention during part production. The need for AM products to be fabricated directly would lead to shorter production lead times, avoid adding or multiplying inaccuracies in indirect processes, and manufacture products that simply cannot be made using conventional techniques. Further development of these techniques has also led to researchers understanding the entire product development cycle on a larger scale and sustainability issues and viewpoints which would enable further commercialization of this technology.

This Handbook on Sustainability in Additive Manufacturing is planned to be published in two volumes and the current one is the first volume which carries six very informative and well-written chapters. Dealing with the core concepts of the subject, these six chapters are written by experts in the field.

Chapter "Sustainable Impact Evaluation of Support Structures in the Production of Extrusion-Based Parts", authored by *Henrique A. Almeida and Mário S. Correia* is concerned with delivering key awareness to the users of extrusion-based systems for a lower environmental impact assessment by means of evaluating the environmental impact of the support production methodologies. In this study, the above-mentioned authors deal with the evaluation (correlating the volume of support material and the time needed for its dissolution) of extra production time involved in the production of the support structures and support structure removal with the aid of two different models with different support material production schemes. The results arising from the study demonstrate that different support production schemes have a significant environmental impact regarding both the production and its dissolution.

Chapter "A New Variant of Genetic Programming in Formulation of Laser Energy Consumption Model of 3D Printing Process", written by *Akhil Garg, Jasmine Siu Lee Lam, and M.M. Savalani* focuses chiefly on the optimization of energy consumption for improving the environmental performance of 3D printing processes. A complexity-based evolutionary approach of genetic programming (CN-GP) in the formulation of functional expression between laser energy consumption, total area of sintering, and two inputs of 3D printing process (selective laser sintering (SLS)) is proposed in this study. The performance of the proposed laser energy consumption models is evaluated against actual experimental data based on five statistical metrics and hypothesis testing, and it is found that slice thickness has 98 % impact on the laser energy consumption in the process.

Chapter "3D Printing Sociocultural Sustainability", authored by *Jennifer Loy, Samuel Canning1, and Natalie Haskell* discusses the potential role of 3D printing in supporting sociocultural sustainability. In this chapter, the impact of digital fabrication on practice for designer/makers is explored in relation to its potential to support the retention of craftsmanship skills, values, and cultural referencing particular to a community involved with craft practice.

Chapter "Additive Manufacturing and its Effect on Sustainable Design", written by *Diegel O, Kristav P, Motte D, and Kianian B* examines various aspects of additive manufacturing from a sustainable design perspective and also looks at the potential to create entirely new business models which could bring about the sustainable design of consumer products. After performing systematic literature studies and examining various case studies, the conclusion is reached that there is a likelihood that additive manufacturing could allow more sustainable products to be developed, but that more quantifiable research is needed in the area to allow designers to exploit better the features of additive manufacturing that can maximize sustainability.

Chapter "Sustainable Design for Additive Manufacturing Through Functionality Integration and Part Consolidation", authored by *Yunlong Tang, Sheng Yang, and Yaoyao Fiona Zhao* deals with a new design philosophy for additive manufacturing with a thorough review of lattice structure design and optimization methods and design for additive manufacturing methods. A general design framework, which has less part counts and less material but without compromising its functionality, to support sustainable design for additive manufacturing via functionality integration and part consolidation is proposed in this chapter. Additionally, a case study is also covered to illustrate and validate the proposed design methodology. This case study concludes that the environmental impact of a product's manufacturing process can be reduced by redesigning the existing product based on the proposed design methodology.

Chapter "Redesigning Production Systems", written by *Jennifer Loy and Peter Tatham* focuses on providing a vision for the redesign of current production systems, supply chains, and values that serves as a starting point for re-establishing the human relationship with manufacturing and business practice. Discussions in this chapter also include the current drivers for change and opportunities for reducing the environmental impact of production systems directly enabled by additive manufacturing.

References

Chua CK, Leong KF (2003) Applications and examples. In rapid prototyping: principles and applications in manufacturing. World Scientific Publishing Company, Singapore. ISBN 9812391201

Gibson I, Dongping S (1997) Material properties fabrication parameters in selective laser sintering process. Rapid Prototyping J 3(4):129–136

Wohlers Report (2013) Additive manufacturing and 3D printing state of the industry annual worldwide progress report. ISBN 0-9754429-9-6

Sustainable Impact Evaluation of Support Structures in the Production of Extrusion-Based Parts

Henrique A. Almeida and Mário S. Correia

Abstract Sustainability creates and maintains the conditions under which humans and nature can exist in a productive harmony, fulfilling the social, economic and other requirements of present and future generations. Environmental and social concerns about human society's impact on the natural environment have been pushing sustainable development issues. Sustainable industrial practices can contribute to the development of more sustainable materials, products, and processes. It is critical to apply eco-design principles and develop greener products and production processes, reducing impacts associated with production and consumption. Bearing this in mind, additive manufacturing has the capability of producing components with the lowest amount of raw material. Alongside with the raw material, in some additive manufacturing systems, support material is needed in order to undergo the production. This present work aims to evaluate the environmental impact of the support production methodologies in order to deliver awareness to the users of extrusion-based systems for a lower environmental impact assessment. The extra production time involved in the production of the support structures and the support structure removal is evaluated. The evaluation consisted of correlating the volume of support material and the time needed for its dissolution. Two different models were then compared with different support material production schemes, regarding the total energy consumption and its environmental impact. The results demonstrate that different support production schemes have significant environmental impact regarding both production and its dissolution.

Keywords Additive manufacturing · Environmental impact · Fused deposition modelling · Support structures · Sustainable manufacturing

H.A. Almeida (✉) · M.S. Correia
School of Technology and Management, Polytechnic Institute of Leiria, Leiria, Portugal
e-mail: henrique.almeida@ipleiria.pt

M.S. Correia
CEMUC, University of Coimbra, Coimbra, Portugal

© Springer Science+Business Media Singapore 2016
S.S. Muthu and M.M. Savalani (eds.), *Handbook of Sustainability in Additive Manufacturing*, Environmental Footprints and Eco-design of Products and Processes, DOI 10.1007/978-981-10-0549-7_2

1 Introduction

Since the start of the industrial revolution, manufacturing processes have shown a rapid and escalating development. Processes and practices have improved, new technologies have been developed and the size and scale of industrial production has expanded enormously, increasing the consumption of both raw materials and energy, in spite of the growing developments in the material field and new sources of energy. According to Gebler et al. (2014), the Industrial Metabolism (the transformation of matter, energy and labour into goods, services, waste and ambient emissions) has generated high levels of economic wealth, simultaneously increasing human interference with the biosphere (Ayres and Simones 1994; Solomon et al. 2007; UNEP 2012). Therefore, cleaner production and environmental sustainability have become a key concern for worldwide government policies, businesses and public in general (Finnvedena et al. 2009). Sustainability is a consideration of resource utilisation without depletion or adverse environmental and ecological impact, minimising the impacts of human actions. In manufacturing, sustainability issues include raw material, energy consumption, waste generation, water consumption, use of environmentally damaging process enablers (e.g., cutting fluids, lubricants, etc.) and the environmental impact of the manufactured part in service (Sreenivasan et al. 2010; Mihelcic et al. 2003). Resource-efficient means of production can contribute to the development of more sustainable products and manufacturing processes (Berry 2004; Howarth and Hadfield 2006), preventing climate change impacts, exhaustion of natural resources and disruptions of ecological systems (Ayres and Simones 1994; Parry et al. 2007).

Among the existing manufacturing technologies, additive manufacturing is an innovative way of producing components and it possesses good environmental characteristics (Gebler et al. 2014; Luo et al. 1999; Beaman et al. 1997). Additive manufacturing has the potential of reducing resource and energy demands as well as process-related CO_2 emissions (Gebler et al. 2014; Kreiger and Pearce 2013; Baumers 2012; Baumers et al. 2011; Campbell et al. 2011; Petrovic et al. 2011). According to Serres et al. (2011), the energy consumed by additive manufacturing to produce parts is also limited when compared to conventional machining processes (Bourhis et al. 2013, 2014). By utilising only the amount of material needed for the building of the final part, additive manufacturing technologies reduces the material mass and energy consumption when compared to conventional subtractive techniques by eliminating scrap, on top of eliminating the need for tooling and the use of environmentally damaging process enablers (Gebler et al. 2014; Sreenivasan et al. 2010; Hague 2005).

A variety of industrial sectors have also embraced additive manufacturing for remanufacturing existing products, an effective approach to reduce simultaneously both costs and environmental impacts, instead of beginning an original production. Additive manufacturing also has the ability to eliminate completely supply chain operations associated with the production of new tooling, enabling the repair and remanufacture of obsolete or failed tools and dies (Sreenivasan et al. 2010; Reeves 2009; Morrow et al. 2007).

Additive manufacturing machines are usually small and therefore can be easily located near to any existing market, thereby reducing the logistics of moving products around the world (Gibson et al. 2015). On the other hand, raw materials for additive manufacturing systems are quite common, which also leads to a large reduction in both transportation costs related to accessibility (Gibson 2011) and the carbon footprint decrease with the consequent reduction of fuel consumption.

Several authors, namely Sreenivasan et al. (2010), Reeves (2009), and Bourell et al. (2009), have defined the carbon footprint reduction of additive manufacturing technologies. According to them, there are five main environmental and sustainable benefits in adopting these technologies:

- More efficient and reduced usage of raw materials required in the supply chain. Hence, reduced need to mine and process primary material ores from our natural resources.
- Displacing of energy-inefficient and wasteful manufacturing processes, such as casting or processes such as CNC machining which requires cutting fluids.
- Ability to design more efficient products with improved operational performances that are more efficient than conventionally manufactured components by incorporating conformal cooling and heating channels and gas flow paths, etc.
- Ability to eliminate fixed asset tooling, allowing for manufacture to occur at any geographic location, such as near to the customer, reducing transportation costs within the supply chain and contributing to diminishing the carbon footprint.
- Lighter weight parts, which when used in transport products such as aircraft increase fuel efficiency and reduce carbon emissions.

Additive manufacturing also presents some sustainability disadvantages. Some processes require support structures that are discarded after each part is built, which in some cases can be equivalent to the amount of part material, if the initial preparation phase isn't properly analysed or even more depending on the complexity of the part being produced (Gibson 2011; Hopkinson et al. 2006). Additive manufacturing machines need a controlled environment without excessive heat and humidity for both machine and raw material, and the energy used to keep the machines working effectively is a negative. In addition, the energy usage for additive manufacturing systems is generally unfavourable (Atkins 2007), as some machines need to use pre-heated and air controlled building chambers, as well the energy required for the processing of raw materials, such as lasers (Gibson et al. 2015; Chua and Leong 2014).

This present work aims to evaluate the ecological impact of the support production methodologies to deliver awareness to the users of extrusion-based systems for a lower environmental impact assessment. The evaluation consists of correlating the volume of support material and the time needed for its dissolution. Two different models are then compared with different support material production schemes, regarding the time involved in the production of the support structures, support dissolution, energy consumption and its environmental impact.

2 Additive Manufacturing

According to Wohlers Associates, additive manufacturing is defined as the process of joining materials to make objects from 3D model data, usually layer upon layer, as opposed to subtractive manufacturing methodologies (Caffrey and Wohlers 2015). The main feature of additive manufacturing, its ability to produce parts of virtually any shape complexity, is huge, as the process is capable of creating mind boggling geometries in spite of their functionality, requirements and materials (Gibson et al. 2015; Hascoet et al. 2014; Hopkinson et al. 2006).

All existing additive manufacturing processes require input data from a 3D digital model, either a solid or surface Computer Aided Design (CAD) model or an existing STereoLithography (STL) file model, which is the current industrial standard for facetted models. In the case of a CAD model, it is tessellated and exported as an STL file so that it may be imported to the manufacturing system's proprietary software. In some additive manufacturing systems, support structures are necessary to embrace overhangs; in this case, the system's proprietary software performs the design of these support structures. The model is then sliced and the sliced data are sent to the additive hardware machine for the production of the final physical part (Gibson et al. 2015; Bártolo et al. 2009).

In addition to the standard importing file, the STL file format, all additive manufacturing systems share another common concern, the part's orientation during production. Part orientation refers to the building direction regarding the part in which the slices are built in the additive manufacturing system (Gibson et al. 2015; Rosen 2014; Allen and Dutta 1995). Determination of the optimal part orientation is a critical issue during the production process in additive manufacturing (Gibson et al. 2015; Rosen 2014; Zhang and Bernard 2013; Alexander et al. 1998), because the building direction has a significant effect on the part's characteristics: such as:

- Dimensional accuracy (Volpato et al. 2014; Saqib and Urbanic 2012; Equbal et al. 2011; Sood et al. 2009, 2010)
- Surface roughness (Vijay et al. 2012, 2014; Armillotta 2006; Onuh and Hon 1998)
- Mechanical properties (Kotliniski 2014; Sood et al. 2012; Majewski and Hopkinson 2011; Quintana et al. 2010; Lee et al. 2007; Ajoku et al. 2006; Ang et al. 2006; Chockalingam et al. 2005; Gibson and Shi 1997)
- Building time and support structures(Gibson et al. 2015; Rosen 2014; Pham and Demov 2001; Chua and Fai 2000; Alexander et al. 1998)
- Cost (Durgun and Ertan 2014; Kumar and Regalla 2011)

Determining the optimal part orientation is both a difficult and a time-consuming task as one has to trade off various contradicting objectives such as part surface finish and building time (Gibson et al. 2015; Rosen 2014; Pham and Demov 2001; Chua and Fai 2000; Alexander et al. 1998). An inadequate choice may result in physical models with a significant "staircase effect" resulting in parts of poor surface quality (Thrimurthulu et al. 2004). Another aspect to be considered is the production of the support structures during production. An inappropriate orientation may result in an excessive production of support structures around the physical part or creation of supports within specific areas of the physical part, which are difficult or almost impossible to remove, increasing significantly the effort and energy for the removal of the support structures (Gibson et al. 2015; Chua and Leong 2014).

According to the ASTM Committee F42 on Additive Manufacturing Technologies, the existing additive manufacturing technologies are classified as follows (ASTM F2792 2015; Gao et al. 2015; Gibson et al. 2015):

1. Material extrusion—process that creates layers by mechanically extruding molten thermoplastic material onto a substrate.
2. Powder bed fusion—these techniques use an energy beam, either a laser or electron beam, to melt selectively a powder bed.
3. Vat photopolymerization—an ultraviolet laser is used to polymerize selectively a UV curable photosensitive resin to create a layer of solidified material. Layers are subsequently cured until the part is complete.
4. Material jetting—processes that directly deposit wax or photopolymer droplets onto a substrate via drop-on-demand inkjetting.
5. Binder jetting—process that deposits a stream of particles of a binder material over the surface of a powder bed, joining particles together where the object is to be formed.
6. Sheet lamination—layers of adhesive-coated paper, plastic or metal laminates are successively glued together and then cut to shape with a knife or laser cutter.
7. Directed energy deposition—metallic powder or wire is fed directly into the focal point of an energy beam to create a molten pool.

According to the previous classification, each additive manufacturing process has its particular process of producing the necessary support structures for embracing the physical parts, along with a method for removing them. Brief descriptions of some of the most relevant processes and how the support structures are produced and then removed are presented in the following sections.

2.1 Material Extrusion

The material extrusion technique, commercially known as Fused Deposition Modelling (FDM), was developed by Crump (1989). Thin crystalline or amorphous thermoplastic filaments are melted by heating and guided by a robotic device controlled by a computer, producing 3D objects (Fig. 1). The model material leaves the extruder in a liquid form and begins to harden (Gibson et al. 2015; Hopkinson et al. 2006). The previously formed layer is the substrate for the next layer and, to assure good interlayer adhesion, the extruded polymer must be maintained at a temperature just below its solidification point. This is possible through

Fig. 1 Illustration of the material extrusion process

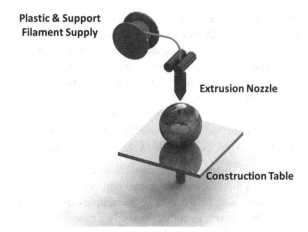

Fig. 2 Physical part with its support material before and after the removal process

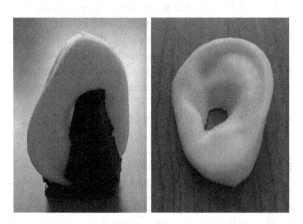

previously heating the building chamber and maintaining its temperature during production (Gibson et al. 2015; Chua et al. 2003).

During the production of the extruded parts, two modeller materials are dispensed through a dual tip mechanism in the extrusion head (Gibson et al. 2015; Hopkinson et al. 2006). As mentioned before, a primary modeller material is used to produce the physical part and a secondary material is used to produce the support structures bonding the primary material of the physical model (Chua et al. 2003). For FDM systems, there are two types of release materials, namely, release materials that can be easily broken off (Break Away Support System) or simply washed away (WaterWorks™ Soluble Support System). Figure 2 illustrates a produced part with its support material before the removal process.

The support structures can be generated either automatically or in more expensive systems, and the user is able to design the best strategy to build the supports structures. To remove these support structures from the parts, even in the case of washable supports, strategy or planning should be taken in account. For instance, features with hollowed out parts, undercuts, constraining features and other interlocking features may be a challenge to achieve complete success in the task of removing the support structures. Currently, as stated earlier, there are two types of support materials, namely, support materials that can be easily broken off or simply washed away.

2.2 Powder Bed Fusion

This process uses an energy beam, either an infrared laser or an electron beam, to heat selectively powder material just beyond its melting point. The laser traces the shape of each cross-section of the model to be built, sintering powder in a thin layer. It also supplies energy that not only fuses neighbouring powder particles, but also bonds each new layer to those previously sintered. Once a layer is scanned, the piston over the model retracts to a new position and the next layer of powder is spread via a rolling mechanism. The powder that remains unaffected by the laser acts as a natural support for the model and remains in place until the model is complete (Fig. 3). Polymer powder bed fusion, namely the Selective Laser Sintering (SLS), which was initially developed by Deckard and Beaman in the mid-1980s, only processed polyamides and polymer composites. Other systems such as Direct Metal Laser Sintering (DMLS), Selective Laser Melting (SLM) and Electron Beam Melting (EBM) were developed later in 1995 and made commercially available in 2005 by EOS GmbH (Germany) and Arcam AB (Sweden), respectively. The actual building process is carried out in a vacuum or inert environment to avoid metal oxidation.

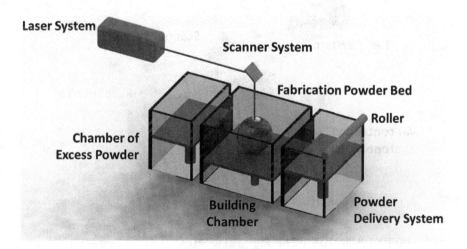

Fig. 3 Illustration of the powder bed fusion process

Regarding the use of support structures during production, there are two differences between the SLS process and the DMLS, SLM and EBM processes. During production, the powder bed serves as a support structure for the model being built, whereas with the other three processes, in spite of the powder bed, the system still produces support structures to sustain the model being built.

2.3 Vat Photopolymerization

In 1984, Charles Hull of 3D System Corp. developed the first stereolithographic apparatus. Stereolithographic processes involve selective polymerization or solidification of liquid photosensitive polymers, namely UV curable resins, through the use of an irradiation UV light source, which supplies the energy needed to induce a chemical reaction, bonding large numbers of small molecules and forming a highly cross-linked polymer. These processes usually employ two distinct methods of irradiation. The first is the mask-based method in which an image is transferred to a liquid polymer by irradiating through a patterned mask. The irradiated part of the liquid polymer is then solidified (Fig. 4). In the second method, a direct writing process using a focused UV beam or laser produces the solid polymer structures.

In the vat photopolymerization process, the support structures are built in the same material as the model part, but in thinner thicknesses of material so that they may be removed manually after concluding the production.

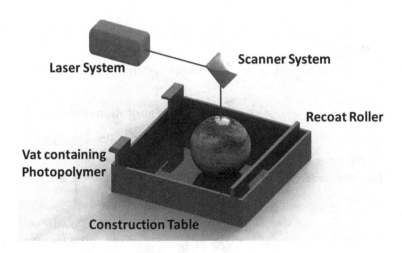

Fig. 4 Illustration of the vat photopolymerization process

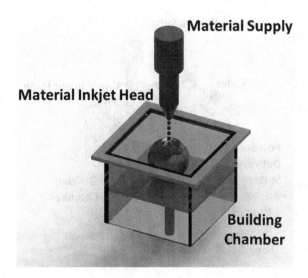

Fig. 5 Illustration of the material jetting process

2.4 Material Jetting

Similar to the normal ink-jet printing technology which transfers ink droplets onto a sheet of paper, material jetting processes directly deposit wax and/or photopolymer droplets onto a substrate via drop-on-demand inkjetting (Fig. 5). The solidification of the jetted droplets occurs via heating or photocuring.

2.5 Binder Jetting

The binder jetting process deposits a stream of particles of a binder material over the surface of a powder bed, joining particles together where the object is to be formed. Recoating occurs via powder spreading as a piston lowers the powder bed so that a new layer of powder can be spread over the surface of the previous layer and then selectively joined to it (Fig. 6). After completing the fabrication process, the unbounded powder material is removed and then part is submitted to an infiltration during post-processing in order to acquire sufficient strength. This method was first studied in MIT and later on commercialized by Z Corporation and ExOne. Similar to the SLS process, the powder bed serves as a support structure during the construction of the part.

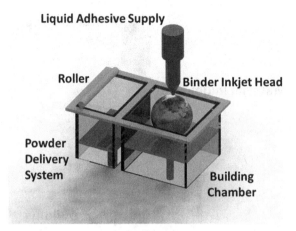

Fig. 6 Illustration of the binder jetting process

2.6 Sheet Lamination

Helisys Inc., now known as Cubic Technologies, commercialize systems using the Laminated Object Manufacturing (LOM) process, which was developed in 1986 and patented in 1987. For the production of the physical parts, this process employs the successive cutting, stacking and gluing of profiled material sheets, either polymer or metallic laminates. The advantages include low internal tension and fragility of the parts, high surface finish details, and lower material, machine and process costs. Disadvantages of this process include the possibility of delamination of the produced part, effort and time involved in decubing the excess material and production of high amounts of waste material.

2.7 Directed Energy Deposition

In these processes, metallic powder or wire is fed directly into the focal point of an energy beam to create a molten pool with the aid of a robotic multi-axis system. In summary, the processes are essentially three-dimensional welding machines. Lasers and electron beams are commonly used as a directed energy source during the process. This process not only allows the production of new metal components but is also used to repair parts, when the damaged portion is reconstructed selectively. Another advantage is the capability of improving tribological performance of any engineered products and the ability to add coatings to existing surfaces. Laser Engineered Net Shaping (LENS) was developed in 1995 at Sandia National Laboratories and is being commercialized by Optomec.

3 Case Study

Our study focussed on one of the existing material extrusion processes in order to evaluate the environmental impact of the energy involved during production and the removal of the support structures. One of the existing FDM systems is the Fortus 450mc 3D Production System from Stratasys (Fig. 7) which enables both parts and models to be built quickly and directly from a CAD STL model. This printer builds three-dimensional parts by extruding a bead of thermoplastic material through a computer-controlled extrusion head, producing high quality parts ready to use immediately after completion. This printer builds models in a wide range of materials: ABS-M30 in six colours for great tensile, impact and flexural strength; ABS-M30i for biocompatibility; ABS-ESD7 for static dissipation; ASA for UV stability and the best aesthetics; PC-ISO for biocompatibility and superior strength; PC for superior mechanical properties and heat resistance; FDM Nylon

Fig. 7 Fortus 450mc 3D production system from Stratasys (Stratasys Ltd. 2015)

12 for maximum toughness; and ULTEM 9085 for high strength-to-weight ratio and favourable FST rating. The Fortus 450mc system builds parts with a maximum size of 406 × 355 × 406 mm.

Regarding the support material, the Fortus 450mc system uses washable and breakaway support materials. In the case of washable release material, this system has four possible support building schemes to support the desired part during production. These four schemes are: Smart, Sparse, Basic and Surround. In each type of support production scheme, not only does the amount of support material vary, so does the global production time of both part and support material. Higher production times mean higher consumption of energy. Afterwards, in the washable support removal tank, greater amounts of support materials increase the time taken for support removal from the final part. Therefore the focus of our study is to evaluate the environmental impact of the support material. The environmental impact value of the release material was not considered because there is no eco-indicator value for the given material. The system supplier only mentions that the support material has an Ecoworks pH level of 12.6, and that it meets most worldwide wastewater requirements (Stratasys Ltd. 2010). Nevertheless, during production, the energy consumption includes the energy for the extrusion of the model and the release material and the energy needed for the support removal tank. Because energy consumption has an eco-indicator impact value, it is considered in this study.

Our research work concerned two steps:

1. The first step was to evaluate and define a relationship between the time necessary to dissolve the support material and the volume of support material
2. The second step considered a case study of two models that were to be produced considering all four schemes of production of support material embracing the desired parts.

3.1 Relationship Between Dissolution Time and Material Volume

As mentioned before, the first step of our research was to evaluate and define a relationship between the time necessary to dissolve the support material and the volume of support material. In order to obtain a relationship between the dissolution time and material volume, three square blocks of different sizes of support material was produced, which would later be dissolved and timed. For each block size, three samples were considered. The blocks had the dimensions, volumes and production times as given in Table 1.

To evaluate the dissolution of the blocks, an experimental set-up was defined as illustrated in Fig. 8. A heated stirring plate was used in order to heat the solution until a temperature of 70 °C. A magnet was used with the stirring plate to provide movement to the solution. During the dissolution process of the support material

Table 1 Block dimensions, volume of support material and production time for each of the blocks

Block dimensions (mm)	Volume of support material (cm^3)	Production time (h:min)
10×10	0.811	00:05
30×30	9.670	00:20
50×50	39.400	00:42

Fig. 8 Experimental setup for the dissolution of the support material blocks

blocks, besides timing the experiment, both the temperature and pH level were constantly being monitored. Regarding the pH level, during all nine experiments, the pH level was maintained within its recommended levels of optimal performance. A 30-L glass flask was used for the dissolution the support material blocks. According to the recommendations provided by the WaterWorks solution, 48.45 g of powder was diluted in 2142 mL of water. Each time a block was dissolved, a new waterworks solution was prepared.

Fig. 9 Illustration of the dissolution process of one of the support material blocks

Table 2 Dissolution and average dissolution times for each sample

Block dimensions (mm)	Volume of support material (cm³)	Sample number #	Dissolution time (min:s)	Average dissolution time (min:s)
10 × 10	0.811	1	21:32	23:38
		2	22:42	
		3	26:40	
30 × 30	9.670	1	30:37	32:45
		2	33:56	
		3	33:43	
50 × 50	39.400	1	42:41	47:56
		2	53:48	
		3	47:21	

Fig. 10 Linear correlation between the support material volume and dissolution times

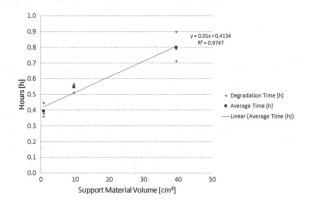

Figure 8 also shows the 50-mm square block, and Fig. 9 illustrates the dissolution of the same block during the experiment. It is possible to observe the interior filling of the block and how the support material was extruded in the block's interior.

Table 2 presents the dissolution times of each sample according to the amount of material of each block and an average dissolution time for each block size. Based on this data, it was possible to define the chart shown in Fig. 10 from which a correlation could be created between the amount of support material and dissolution time.

(a) **(b)**

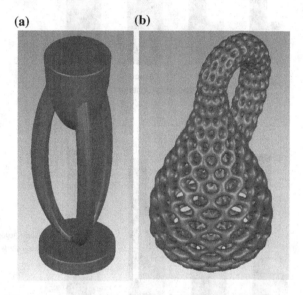

Fig. 11 CAD STL models of **a** Super Rugby Trophy 2015 and **b** Klein Bottle

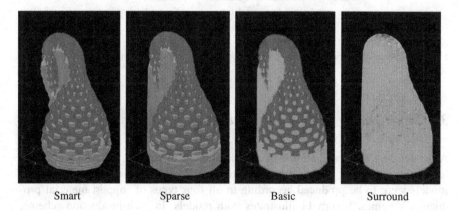

Smart   Sparse   Basic   Surround

Fig. 12 Types of support material production schemes

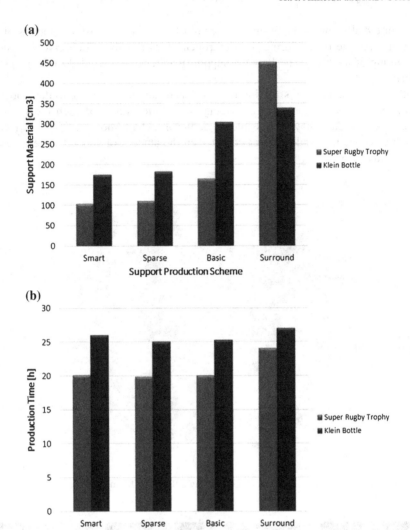

Fig. 13 Amount of **a** support material and **b** production times for each support material production scheme

3.2 Super Rugby Trophy 2015 and Klein Bottle

After defining the correlation between the amount of support material and dissolution time, the next step of our research focused on two case studies where the
models were to be produced according to all four types of support material production schemes. Figure 11 illustrates both models. In each production scheme,
the parts were placed and oriented in the same position. At this stage, we only

Table 3 Amount of model material, support material and production times for each of the support material production schemes

Support production scheme	Model material (cm³)	Support material (cm³)	Production time (h:min)
Super Rugby Trophy 2015			
Smart	625.120	103.200	20:05
Sparse	625.100	109.560	19:50
Basic	626.130	165.130	20:00
Surround	628.200	451.770	24:02
Klein Bottle			
Smart	277.810	174.720	26:01
Sparse	278.070	182.550	25:01
Basic	280.010	303.520	25:13
Surround	280.770	338.950	26:59

accessed the Insight 10.4 software and simulated the production of both parts. Figure 12 illustrates the four types of support production schemes considered, namely: Smart, Sparse, Basic and Surround.

It possible to observe from Fig. 12 that the amount of support material around the model is different in each case, whereas the Surround scheme presents the highest amount of support material. Figure 13 illustrates the difference in amount of support material along with the production time for each type of support production scheme. In terms of the amount of support material, both the Basic and Surround present the highest amount of material needed in the process. Regarding the production time, all four support production schemes present similar production times, except for the Surround which presents the highest time (Table 3).

From the technical specifications of the Fortus 450mc system, it was possible to calculate the energy consumption and respective environmental energy consumption impact. Because the energy consumption and the environmental energy consumption impact are time dependent, the variations observed in the chart of the production times are also observed in these two charts. In other words, for the Smart, Sparse and Basic, in spite of slight variations, the variation might not be considered relevant enough, but for the Surround scheme the difference is relevant enough for discussion (Fig. 14).

Based on the support material dissolution experiment, it is possible to determine the dissolution times for each of the support material production schemes for each model (Fig. 15a). Once the dissolution times were determined, the energy consumption and energy consumption impact for the support material removal was determined (Fig. 15b, c).

Considering that the entire production cycle is composed of both the extrusion and support removal process, Table 4 presents the total values of the production times, energy consumption and energy consumption impacts. The Energy Consumption Impact value is calculated by multiplying the Electricity indicator value of 47 (Electricity Low Voltage Portugal) by the amount of consumed energy

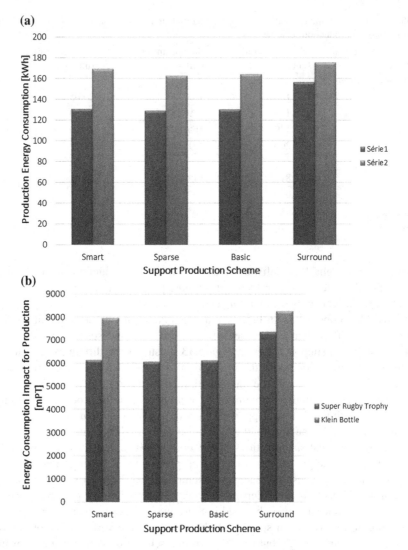

(a)

(b)

Fig. 14 Amount of **a** production energy consumption and **b** production energy consumption impact for each support material production scheme

during the production process and support structure removal (Lofthouse 2006; Goedkoop and Spriensama 2001; Ministry of Housing, Spatial Planning and the Environment 2000).

From the data presented, it is possible to observe that there is a difference of 7.622 h (production time), 34.144 kWh (energy consumption) and 1604.778 mPt (energy consumption impact) between the Surround and Sparse support material production scheme for the Super Rugby Trophy model. Regarding the Klein model, the difference is 3.530 h, 15.911 kWh and 747.833 mPt between the

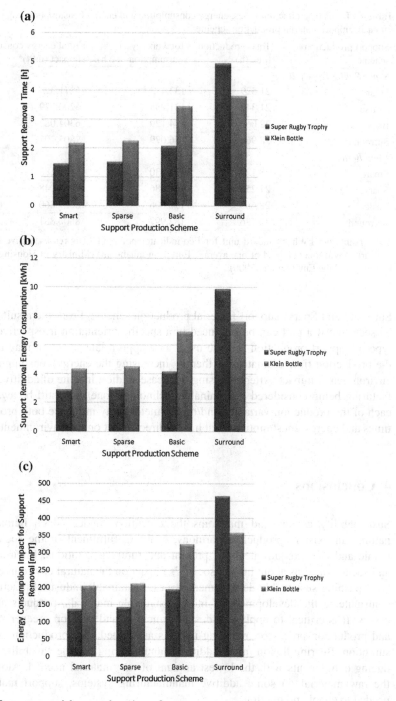

Fig. 15 Support material removal: **a** times, **b** energy consumption and **c** energy consumption impact for each support material production scheme

Table 4 Totals of production times, energy consumption and energy consumption impact values for each support material production scheme

Support production scheme	Total production time (h)	Total energy consumption (kWh)	Total energy consumption impact (mPt)[a]
Super Rugby Trophy 2015			
Smart	21.529	133.432	6271.326
Sparse	21.342	131.935	6200.929
Basic	22.065	134.129	6304.082
Surround	28.964	166.079	7805.707
Klein Bottle			
Smart	28.177	173.430	8151.188
Sparse	27.256	167.086	7853.048
Basic	28.665	170.806	8027.860
Surround	30.786	182.997	8600.881

[a]Milli-points per kWh—standard unit for Eco-indicator points (1 Pt is representative for 1000 of yearly environmental load of one average European inhabitant) (Ministry of Housing, Spatial Planning and the Environment 2000)

Surround and Sparse support material production scheme. From the results, it can be seen that if a part can be produced in a specific orientation irrespective of the type of support production scheme used, an inappropriate selection may increase the production time dramatically, thereby increasing the energy consumption and environmental impact. After analysing both case studies, in spite of additive manufacturing being considered a sustainable technology, one must still be aware that each of the production variables and/or parameters may influence both production times and energy consumptions that have a direct effect on the environment.

4 Conclusions

Sustainability creates and maintains the conditions under which humans and nature can exist in productive harmony, enabling fulfilment of the social, economic and other requirements of present and future generations. Environmental and social concerns about human society's impact on the natural environment have been pushing sustainable development issues. Sustainable industrial practices can contribute to the development of more sustainable materials, products and processes. It is critical to apply eco-design principles and develop greener products and production processes, reducing impacts associated with production and consumption. Bearing this in mind, additive manufacturing has the capability of producing components with the lowest amount of raw material needed. Along with the raw material, in some additive manufacturing systems, support material is needed to enable production.

This present work aims to evaluate the ecological impact of support production methodologies to make users of aware extrusion-based systems for a lower environmental impact assessment. The extra time involved in the production of both models between the Surround and Sparse support material production scheme totals 11.152 h, 50.055 kWh and 2352.611 mPt with respect to the production time, energy consumption and energy consumption impact, respectively. After analysing both case studies, in spite of additive manufacturing being considered a sustainable technology, one must still be aware that each of the production variables and/or parameters may influence both production times and energy consumptions that have a direct effect on the environment.

References

Ajoku U, Saleh N, Hopkinson N, Hague R, Poonjolai E (2006) Investigating mechanical anisotropy and end-of vector effect in laser-sintered nylon parts. Proceedings of the Institution of Mechanical Engineers Part B Journal of Engineering Manufacture 220(7):1077–1086

Alexander P, Allen S, Dutta D (1998) Part orientation and build cost determination in layered manufacturing. Comput Aided Des 30(5):343–356

Allen S, Dutta D (1995) On the computation of part orientation using support structures in layered manufacturing. Journal of Design and Manufacturing 5:153–162

Ang KC, Leong KF, Chua CK, Chandrasekaran M (2006) Investigation of the mechanical properties and porosity relationships in fused deposition modelling-fabricated porous structures. Rapid Prototyping Journal 12(2):100–105

Armillotta A (2006) Assessment of surface quality on textured FDM prototypes. Rapid Prototyping Journal 12(1):35–41

ASTM Standard F2792 (2015) "Standard Terminology for Additive Manufacturing Technologies" URL: http://www.astm.org/Standards/F2792.htm

Atkins (2007) ATKINS Manufacturing a Low Carbon Footprint, ATKINS project report (http://www.atkins-project.com/pdf/ATKINSfeasibilitystudy.pdf)

Ayres RU, Simones UE (1994) Industrial Metabolism— Restructuring for Sustainable Development. The United Nations University, Tokyo/Japan

Bártolo PJ, Almeida HA, Laoui T (2009) Rapid Prototyping & Manufacturing for Tissue Engineering Scaffolds. Computer Applications in Technology 36(1):1–9

Beaman JJ, Barlow JW, Bourell DL, Crawford RH, Marcus HL, McAlea KP (1997) Solid freeform fabrication: a new direction in manufacturing. Kluwer Academic Press, Boston

Baumers M (2012) Economic Aspects of Additive Manufacturing: benefits, Costs and Energy Consumption (Doctoral Thesis) Loughborough University, Leicestershire, United Kingdom

Baumers M, Tuck C, Wildman R, Ashcroft I, Hague R (2011) Energy inputs to additive manufacturing: does capacity utilization matter? In: Conference Paper: Solid Freeform Fabrication Symposium 2011, Austin (TX)/USA

Berry M (2004) The Importance of sustainable development. Canada, Columbia Spectator

Bourell DL, Leu MC, Rosen DW (2009) Roadmap for additive manufacturing: identifying the future of freeform processing. The University of Texas at Austin

Bourhis FL, Kerbrat O, Dembinski L, Hascoet JY, Mognol P (2013) Sustainable manufacturing: evaluation and modelling of environmental impacts in additive manufacturing. Int J Adv Manuf Technol 69:1927–1939

Bourhis FL, Kerbrat O, Dembinski L, Hascoet JY, Mognol P (2014) Predictive model for environmental assessment in additive manufacturing process. Procedia CIRP 15:26–31

Caffrey T, Wohlers T (2015) Wohlers Report 2015—Additive Manufacturing and 3D Printing State of the Industry—Annual Worldwide Progress Report, Wohlers Associates, Inc. ISBN: 978-0-9913332-1-9

Campbell T, Williams C, Ivanova O, Garrett B (2011) Could 3D printing change the world? Technologies, and implications of additive manufacturing. Atlantic Council, Washington DC/USA

Chockalingam K, Jawahar N, Ramanathan KN, Banerjee PS (2005) Optimization of stereolithography process parameters for part strength using design of experiments. Int J Adv Manuf Technol 29(1–2):79–88

Chua CK, Fai LK (2000) Rapid prototyping: principles and applications in manufacturing. World Scientific

Chua CK, Leong KF (2014) 3D printing and additive manufacturing—principles and applications, 4th edn. World Scientific Publishing

Chua CK, Leong KF, Lim CS (2003) Rapid prototyping: principles and applications in manufacturing, 2 edn. World Scientific

Crump SS (1989) Apparatus and method for creating three-dimensional objects, US Pat. 5121329

Durgun I, Ertan R (2014) Experimental investigation of FDM process for improvement of mechanical properties and production cost. Rapid Prototyping J 20:228–235

Equbal A, Ohdar RK, Mahapatra SS (2011) Prediction of dimensional accuracy in fused deposition modelling: A fuzzy logic approach. Int J Productivity Qual Manage 7(1):22–43

Finnvedena G, Hauschildb M, Ekvallc T, Guinée J, Heijungs R, Hellweg S, Koehler A, Pennington D, Suh S (2009) Recent developments in Life Cycle. J Environ Manage 92(1):1–21

Gao W, Zhang Y, Ramanujan D, Ramani K, Chen Y, Williams CB, Wang CCL, Shin YC, Zhang S, Zavattieri PD (2015) The status, challenges, and future of additive manufacturing in engineering. Comput Aided Des 69:65–89

Gebler M, Uiterkamp AJMS, Visser C (2014) A global sustainability perspective on 3D printing technologies. Energy Policy 74:158–167

Gibson I (2011) Is additive manufacturing a sustainable technology? In: Bártolo H et al (eds) Proceedings of SIM2011 Sustainable Intelligent Manufacturing. IST Press, pp 583–589

Gibson I, Shi D (1997) Material properties and fabrication parameters in selective laser sintering process. Rapid Prototyping J 3:129–136

Gibson I, Rosen D, Stucker B (2015) Additive manufacturing technologies—3D printing, rapid prototyping and direct digital manufacturing, 2nd edn. Springer, New York

Goedkoop M, Spriensama R (2001) The eco-indicator 99: a damage oriented method for Life Cycle Impact Assessment. Methodology Report, Netherlands

Hague R (2005) Unlocking the design potential of rapid manufacturing. In: Hopkinson N et al (eds) Rapid manufacturing: an industrial revolution for the digital age. Wiley

Hascoet JY, Marya S, Marya M, Singh V (2014) Materials science challenges in the additive manufacturing of industrial parts. In: Kai CC et al (eds) Proceedings of the 1st International Conference on Progress in Additive Manufacturing (Pro-AM2014). Research Publishing Services, pp 133–138

Hopkinson N, Hague RJM, Dickens PM (2006) Rapid Manufacturing—an industrial revolution for the digital age. Wiley, England

Howarth G, Hadfield M (2006) A sustainable product design model. Mater Des 27(10):1128–1133

Kotliniski J (2014) Mechanical properties of commercial rapid prototyping materials. Rapid Prototyping J 20(6):499–510

Kreiger M, Pearce JM (2013) Environmental life cycle analysis of distributed three-dimensional printing and conventional manufacturing of polymer products. ACS Sustain Chem Eng 1(12):1511–1519

Kumar GP, Regalla SP (2011) Optimization of support material and build time in Fused Deposition Modeling (FDM). Appl Mech Mater 110–116:2245–2251

Lee CS, Kim SG, Kim HJ, Ahn SH (2007) Measurement of anisotropic compressive strength of rapid prototyping parts. J Mater Process Technol 187–188:627–630

Lofthouse V (2006) Ecodesign tools for designers: defining the requirements. J Clean Prod 14(15–16):1386–1395

Luo Y, Ji Z, Leu MC, Caudill R (1999) Environmental performance analysis of solid free-form fabrication processes. In: Proceedings of the 1999 IEEE International Symposium on Electronics and the Environment (ISEE-1999), IEEE

Majewski C, Hopkinson N (2011) Effect of section thickness and build orientation on tensile properties and material characteristics of laser sintered nylon-12 parts. Rapid Prototyping J 17(3):176–180

Mihelcic JR, Crittenden JC, Small MJ, Shonnard DR, Hokanson DR, Zhang Q, Chen H, Sorby SA, James VU, Sutherland JW, Schnoor JL (2003) Sustainability science and engineering: the emergence of a new Metadiscipline. Environ Sci Technol 37(23):5314–5324

Ministry of Housing, Spatial Planning and the Environment (2000) Eco-indicator 99 Manual for Designers—a damage oriented method for life cycle impact assessment. Ministry of Housing, Spatial Planning and the Environment, The Netherlands

Morrow WR, Qi H, Kim I, Mazumder J, Skerlos SJ (2007) Environmental aspects of laser-based and conventional tool and die manufacturing. J Clean Prod 15:932–943

Onuh SO, Hon KKB (1998) Optimising build parameters for improved surface finish in stereo-lithography. Int J Mach Tools Manuf 38(4):329–342

Parry ML, Canziani OF, Palutikof JP, Linden PJ, Hanson JE (2007) Technical summary. Climate change 2007: impacts, adaptations and vulnerability. Contribution of Working Group II to the Fourth Assessment Report of the Intergovernmental Panel on Climate Change. Cambridge University Press, Cambridge, UK, pp 23–78

Petrovic V, Gonzales JVH, Ferrado OJ, Gordillo JD, Puchades JRB, Ginan LP (2011) Additive layered manufacturing: sectors of industrial application shown through case studies. Int J Prod Res 49(4):1071–1079

Pham DT, Demov SS (2001) Rapid manufacturing: the technologies and applications of rapid prototyping and rapid tooling. Springer, London Limited

Quintana R, Choi JW, Puebla K, Wicker R (2010) Effects of build orientation on tensile strength for stereolithography-manufactured ASTM D-638 type I specimens. Int J Adv Manuf Technol 46:201–215

Reeves P (2009) Additive Manufacturing—A supply chain wide response to economic uncertainty and environmental sustainability. International Conference on Industrial Tools and Material Processing Technologies, Ljubljana, Slovenia

Rosen D (2014) What are the principles for design for additive manufacturing? In: Kai CC et al (eds) Proceedings of the 1st International Conference on Progress in Additive Manufacturing (Pro-AM2014), Research Publishing Services, pp 85–90

Saqib S, Urbanic J (2012) An Experimental Study to Determine Geometric and Dimensional Accuracy Impact Factors for Fused Deposition Modelled Parts. In: ElMaraghy HA (ed) Enabling Manufacturing Competitiveness and Economic Sustainability. Springer, Berlin, Heidelberg, pp 293–298

Serres N, Tidu D, Sankare S, Hlawka F (2011) Environmental comparison of MESO-CLAD® process and conventional machining implementing life cycle assessment. J Clean Prod 19(9–10):1117–1124

Solomon S, Qin D, Manning M, Chen Z, Marquis M, Averyt KB, Tignor M, Miller HL (2007) Technical summary: climate change 2007: the Physical Science Basis. Contribution of Working Group I to the Fourth Assessment Report of the Intergovernmental Panel on Climate Change. Cambridge University Press, Cambridge, UK and New York, NY

Sood AK, Ohdar RK, Mahapatra SS (2009) Improving dimensional accuracy of Fused Deposition Modelling processed part using grey Taguchi method. Mater Des 30(10):4243–4252

Sood AK, Ohdar RK, Mahapatra SS (2010) Parametric appraisal of mechanical property of fused deposition modelling processed parts. Mater Des 31(1):287–295

Sood AK, Ohdar RK, Mahapatra SS (2012) Experimental investigation and empirical modelling of FDM process for compressive strength improvement. J Adv Res 3(1):81–90

Sreenivasan R, Goel A, Bourell DL (2010) "Sustainability issues in laser-based additive manufacturing", LANE 2010. Phy Procedia 5:81–90

Stratasys Ltd. (2010) uPrint® and uPrint® Plus Personal 3D Printers User Guide. Stratasys Ltd

Stratasys Ltd. (2015) Fortus 380MC and 450MC Spec Sheet. Stratasys Ltd

Thrimurthulu K, Pandey PM, Reddy NV (2004) Optimum part deposition orientation in fused deposition modeling. Int J Mach Tools Manuf 44:585–594

UNEP (2012) Annual Report 2012. United Nations Environment Programme, Nairobi, Kenya

Vijay P, Danaiah P, Rajesh KVD (2012) Critical parameters effecting the rapid prototyping surface finish. J Mech Eng Autom 1(1):17–20

Vijay I, Chockalingam K, Kailasanathan C, Sivabharathy M (2014) Optimization of Surface roughness in selective laser sintered stainless steel parts. Int J ChemTech Res 6(5):2993–2999

Volpato N, Foggiatto JA, Schwarz DC (2014) The influence of support base on FDM accuracy in Z. Rapid Prototyping J 20:2

Zhang Y, Bernard A (2013) Using AM feature and multi-attribute decision making to orientate part in additive manufacturing. In: Bártolo PJ et al (eds) High value manufacturing. CRC Press, pp 411–416

A New Variant of Genetic Programming in Formulation of Laser Energy Consumption Model of 3D Printing Process

Akhil Garg, Jasmine Siu Lee Lam and M.M. Savalani

Abstract Literature studies reveal that significant work has been done in improving the productivity of the 3D printing process, at the same time neglecting the associated environmental implications. Growing demand for customized and better product quality has resulted in an increase in energy consumption, which is one of the important factors for sub-standard environmental performance. Consequences include the adverse impacts on humans, plant life, and soil and among others. Thus, an optimization of energy consumption is needed for improving the environmental performance of the 3D printing process. In this context, the present work proposes a complexity-based-evolutionary approach of genetic programming (CN-GP) in formulation of functional expression between laser energy consumption, total area of sintering, and two inputs of 3D printing process [selective laser sintering (SLS)]. The performance of the proposed laser energy consumption models is evaluated against actual experimental data based on five statistical metrics and hypothesis testing. Relationships between laser energy consumption and two inputs are unveiled which can be used for effectively monitoring the environmental performance of the SLS process. It was found that the slice thickness has 98 % impact on the laser energy consumption in the process. A major contribution of the study is that the optimum values of inputs can be selected to optimize the energy consumption of the SLS process.

Keywords 3D printing · Energy efficiency · Environmental performance · Genetic programming · Laser energy consumption

A. Garg · J.S.L. Lam (✉)
School of Civil and Environmental Engineering, Nanyang Technological University,
50 Nanyang Ave, Singapore 639798, Singapore
e-mail: sllam@ntu.edu.sg

M.M. Savalani
Department of Industrial and Systems Engineering, Hong Kong Polytechnic University,
Kowloon, Hong Kong

© Springer Science+Business Media Singapore 2016
S.S. Muthu and M.M. Savalani (eds.), *Handbook of Sustainability
in Additive Manufacturing*, Environmental Footprints and Eco-design
of Products and Processes, DOI 10.1007/978-981-10-0549-7_3

1 Introduction

Among 3D printing processes [fused deposition modeling (FDM), selective laser sintering (SLS), selective laser melting and stereolithography (SLA)], SLS has attracted attention because it produces the functional prototypes directly from computer aided design data without much need of human intervention and tools. The process deploys a laser beam to fuse the powder selectively into a designed solid object layer by layer (Deckard and McClure 1988; Raghunath and Pandey 2007; Cong-Zhong et al. 2009). It is also able to produce functional parts from materials such as nylon and titanium. The literature reveals that the characteristics of SLS fabricated parts, such as porosity strength, density, and shrinkage ratio, exhibit high dependence on parameters such as the material and powder properties and other machine specifications (laser power, scan speed, and scan spacing). These characteristics can be improved by an appropriate setting of the input parameters (Nelson et al. 1993; Tontowi and Childs 2001; Cervera and Lombera 1999; Shen et al. 2004; Singh and Prakash 2010; Garg and Lam 2015a; Garg et al. 2015a).

Despite having unusual advantages, SLS is an extensive energy-consuming process and is considered energy inefficient. Mass production of functional components by the SLS process contributes to higher economic growth but the need for energy and materials also grows exponentially, which is not considered environmentally friendly (Paul and Anand 2012). Reducing the energy consumption has become a top priority for industries and government in providing a clean and healthy environment for citizens. A survey study conducted on applications of empirical modeling of 3D printing processes by Garg et al. (2014a) reveals that most attention was paid to improving the product quality/productivity of 3D printing processes (Yang et al. 2002; Chatterjee et al. 2003; Kruth and Kumar 2005; Liao and Shie 2007; Beal et al. 2009; Savalani et al. 2012). Some of the work on the environmental issues in supply chains and logistics has been conducted and solved using multi-criteria decision methods such as Analytic Network Process (Lam 2015; Lam and Lai 2015; Lam and Dai 2015). To the best of the authors' knowledge, hardly any emphasis has been paid to improving the environmental performance and the reduction of energy consumption of 3D printing processes (Paul and Anand 2012) (Fig. 1).

It is known from the literature that the total energy consumed to manufacture the component from the SLS process is derived from energy spent in running the laser systems, part and powder platforms, powder spreading roller, and heating of bed (Paul and Anand 2012) (Fig. 2). Among these sources, the energy consumed in running the laser system is the primary focus because the energy consumed during this phase depends on the machine being used, operating conditions, geometry of component to be fabricated, the value of slice thickness, and the values for orientation chosen for the part build. Numerous studies quantifying the contribution of laser energy to the total energy for the SLS process have been conducted (Mognol et al. 2006; Sreenivasan and Bourell 2009; Kellens et al. 2010; Baumers et al. 2011; Niino et al. 2011). Among these studies, the highest contribution of laser

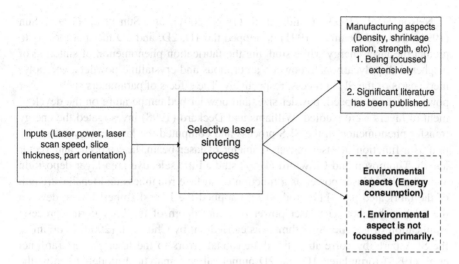

Fig. 1 Manufacturing and environmental aspects of the SLS process

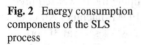

Fig. 2 Energy consumption components of the SLS process

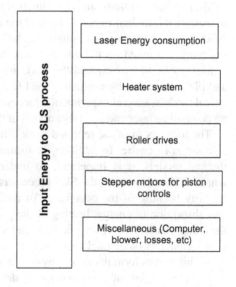

energy to the total energy consumed was found to be 66 % for the experiments conducted on SLS EOS EOSINT M250 Xtended machine (Mognol et al. 2006). However, the minimum contribution of laser energy to the total energy consumed was found to be as low as 1 % for the experiments conducted on the large plastic sintering machines (Semplice, ASPECT) (Niino et al. 2011). The greater contribution of laser energy to the total energy consumed is found to be in the case of machines with smaller build platforms because the energy required for heating the powder bed and moving the build platform decreases.

Nelson et al. (1993), Childs et al. (1999, 2001), and Sun et al. (1990), Sun (1991), Sun and Beaman (1991) developed the 1D, 2D, and 3D thermal models to predict the laser energy while studying the fabrication phenomenon of sintering of bisphenol-A polycarbonate powder, amorphous and crystalline powders, and polymeric parts of the SLS process, respectively. The effects of parameters such as laser power, laser scan speed, powder size, and powder bed temperature on the development of layers were studied. Williams and Deckard (1998) investigated the energy transfer phenomenon in the SLS process and computed the laser energy consumption as a function of laser power, diameter of laser beam, laser speed, and vector length. Thompson and Crawford (1997) studied the selective area layer deposition and formulated laser power as a function of surface roughness and tensile strength of the fabricated part. Fuh et al. (1995) applied the Beer–Lambert law to develop the relationship between laser power and cure depth of the laser curing process. Application of genetic algorithms was carried out by Cho et al. (2000) to optimize the SLA parts by correlating the dimensional errors to the laser power. Yardimci et al. (1995) formulated 1D and 2D numerical and analytical models to solve the 3D energy equation for the deposition and solidification phenomenon in FDM process. Bellini et al. (2004) used three methods, namely power law for Newtonian fluids, transfer functions, and experimental methods, to compute the power required to extrude the molten material through the liquefier of an FDM machine. Xu et al. (2000) analyzed the surface roughness, building time, and cost of several 3D printing processes based on the power source needed to fabricate the component. Luo et al. (1999a, b) investigated the environmental impacts arising from SLS, SLA, and FDM processes by estimating the life cycle energy utilization. Besides the analytical models, several experimental pieces of work have been carried out that focus on computing laser energy consumption in the SLS process (1999).

The findings from the reported work differ from machine to machine and therefore are not generic. In addition, in formulation of the functional relationship of thermal models, it is imperative to understand the process behavior (Paul and Anand 2012). However, the SLS process is characterized by complexity and nonlinearity because of the occurrence of multiple phenomena, such as transmission and absorption of energy, heating of the powder bed, and sintering and cooling of the components. Thus it is difficult to understand the nature of the effect of the process parameters (slice thickness and part orientation) on the laser energy consumption. This makes formulation of physics-based models difficult (Garg et al. 2014b).

If we can think of a mechanism to drive the formulation of models from the given data, this could perhaps be an interesting mode of understanding the hidden principles behind the process on using these models. To meet this objective, a well-known evolutionary approach of genetic programming (GP) can be applied (Koza 1994). Several applications and advancements of GP in the field of 3D printing processes have been conducted. The effects of the process parameters such as laser power, layer thickness, and feed rate on the bead width of a selective laser melting (SLM) fabricated prototype have been investigated by a modified approach of multi-gene genetic programming (Garg et al. 2014c). It was found that the scanning speed has the highest impact on the bead width, followed by laser power and layer thickness. The bead width increases with increase in laser power whereas the other parameters are kept constant. Another application is the formulation of model for density characteristic of the SLS printed parts based on finite element analysis-evolutionary algorithm approach (Vijayaraghavan et al. 2014). In this work, the time-dependent temperature distribution was evaluated for the printed part and density was computed based on varying laser power, layer thickness, and feed rate. The simulated data were fed into a framework of GP and the functional density model was formulated. It was found that the density of the SLS part behaves nonlinearly with laser power and scan velocity whereas it increases with increase in feed rate. Extended application of GP in formulation of the open porosity model was conducted by Garg et al. (2015c). The effect of the process parameters such as layer thickness, laser power, and feed rate on the open porosity of the e-hydroxyapatite (HA)-polyamide (PA) composite was investigated on the SLM experimental set up. This was followed by the application of evolutionary approach of GP in formulation of the open porosity model. The sensitivity and parametric analysis performed on the model revealed that the open porosity behaves nonlinearly with the laser power. This brief review of GP in the modeling of 3D printing processes revealed that they are nonlinear in nature with complex relationships between process parameters, some form of modification in the algorithm framework is needed to improve its extrapolation ability, and none of these studies addressed the actual laser energy consumption in the process. The effective functioning of the GP can be further improved if the definition of complexity of the models during the evolutionary stages of GP (Garg et al. 2014a, b, c) is appropriately defined. Complexity of the evolved models during the evolutionary stages of GP is generally defined by the number of nodes of the tree. This implies that $\sin(x)$ and $-x$ have the same complexity as they both have two nodes, but it is not at all true. This is a critical issue because the complexity term is a component of the fitness function which monitors the evolutionary search and the convergence rate towards achieving the optimum solution. Therefore, determining its correct value is essential for the effective functioning of the algorithm by driving the evolution to its direction of global minimum. This issue also requires a thorough investigation and therefore forms a motivation for authors in developing a framework that can result in evolution of generalized models in effectively studying the impact of input process parameters on the laser energy consumption.

In this work, a new variant of GP, i.e., complexity measure-based evolutionary framework of GP (CN-GP), is proposed to formulate the relationship between the total area of sintering (TAS), laser energy consumption, and input process parameters (slice thickness and part orientation) of the SLS process. The procedure for modeling the laser energy consumption of the SLS process is shown in Fig. 3. One objective of the present work is to explore the ability of CN-GP in formulation of generalized functional relationship of TAS and laser energy consumption of an SLS fabricated prototype. Experiments on SLS are conducted with the measurement of TAS and laser energy consumption based on the two input process parameters (slice thickness and build part orientation). This is followed

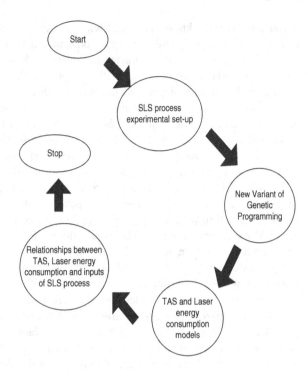

by the application of the two methods (proposed and standardized GP) to the data
obtained from the experiments. The performance of the models formulated from
these two methods is evaluated on the actual experimental data using the five statistical metrics and hypothesis testing. Furthermore, sensitivity and parametric
analysis is conducted to validate the robustness of the proposed models by unveiling the dominant input variables and the hidden nonlinear relationships between
energy consumption and input parameters. The information extracted from the
relationships is useful for monitoring the energy component of the SLS process
which can then contribute to a greener environment.

2 Experiment Details for the SLS Process in the Measurement of Laser Energy Consumption and TAS

This section describes the experimental set-up for the measurement of total area of
sintering (TAS) and laser energy consumption. Laser energy consumption is estimated by virtual manufacturing of a component and correlating to slice thickness

Table 1 Values chosen for input process parameters for measuring TAS and laser energy consumption

Input process parameters	Level 1	Level 2	Level 3	Level 4	Level 5	Level 6
Slice thickness (mm)	0.03	0.05	0.06	0.07	0.09	0.1
Built part orientation (degrees)	0	15	22	30	37	45

and part orientation. Absorptivity of the laser power system is assumed to be 0.95 (Nancharaiah et al. 2013). The other parameters, such as laser power of 70.00 W, beam radius of 17.50 μm, and laser scan speed of 1 m/s, are chosen from Nelson et al. (1993) from the 3D Systems Sinterstation Pro machine. The step-by-step procedure for computing the laser energy consumption in SLS process is as follows:

1. First, the part to be built is modeled in CAD and then exported in STL format.
2. The values for slice thickness and part orientation are chosen. Table 1 shows the values selected for each slice thickness and part orientation.
3. The STL file is then sliced and the sintering area for each slice is calculated by using the connect hull approach.
4. TAS is calculated by summing the areas for each slice.
5. From TAS, the laser energy consumption is calculated.
6. Repeat steps 2–5 for different values for slice thickness and built part orientation.

Full factorial experiments with orthogonal arrays are used for conducting a number of experimental runs in measuring the TAS and laser energy consumption in the process. A total of 40 samples were produced by SLS machine for different values of slice thickness and part orientation (Table 1) with dimensions $10 \times 5 \times 1.5$.

For each slice part, the sintering area is calculated and then summation is done for all slices to compute the TAS. From this TAS, further laser energy consumption is calculated. Findings from the experiments reveal that the laser energy decreases with increase in slice thickness whereas for part orientation it first increases and then decreases.

For training of the data obtained from the experiment discussed (Table 2), the two inputs (slice thickness and part orientation) and two outputs (TAS and laser energy consumption) are fed into the cluster of two variants of GP. To take into account the concern of selection of training and testing data set that may affect the training of the models, 80 % (32 samples) of the collected data was randomly chosen as training samples with the remainder as testing data. The training data were used for formulating the models with the test data samples used for testing the generalization ability of the models. In the following section, the proposed variant of GP is briefly discussed.

Table 2 Experimental runs conducted for SLS process in measurement of TAS and laser energy consumption

Trial No.	Slice thickness (mm) (x_1)	Build part orientation (degrees) (x_2)	Total area of sintering (TAS) (mm^2)	Laser energy consumption (kilojoules)
1	0.03	0	7200	17.41
2	0.04	15	7716	10.08
3	0.05	22	4318	10.49
4	0.06	30	2150	10.54
5	0.07	37	2798	9.87
–	–	–	–	–
–	–	–	–	–
–	–	–	–	–
–	–	–	–	–
38	0.04	45	7674	10.26
39	0.05	30	4367	10.55
40	0.09	22	7683	15.89

3 Evolutionary Algorithms

3.1 Complexity-Based Evolutionary Approach of Genetic Programming (CN-GP)

The difference between the complexity measure-based framework of GP (CN-GP) and the standardized one (Koza 1994) is the addition of a new complexity measure in its fitness function. The proposed framework (Fig. 4) is described in three subcategories as follows:

1. *Initialization*
   In the first step, the functional and terminal set is defined. Operational elements (arithmetic operators $(+, -, /, \times)$, nonlinear functions (sin, cos, tan, exp, tanh, log), or Boolean operators) form the functional set. Variable and constant elements (input variables such as suction and stress, the range of random constants) form the terminal set. The range of random constants chosen is -10 to $+10$. The elements from these two sets are combined randomly to form a model. In this way, several models are evolved.
2. *Evaluation of performance of models based on definition of complexity term in fitness function by orthogonal polynomials*
   The performance of the individuals in the initial population is evaluated based on the fitness function [structural minimization principle (SRM)] so as to avoid *overfitting*. Polynomials of order 1–6 are used to define the complexity of the models evolved during the evolutionary stage. The minimum order of the polynomial that best fits the GP model is considered as the complexity of that model.

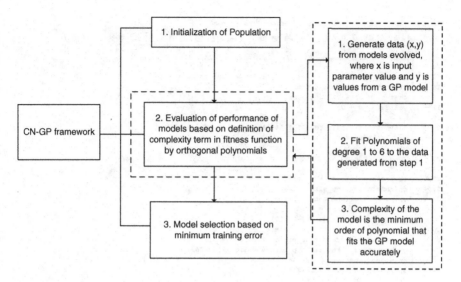

Fig. 4 Flowchart showing step-by-step procedure of CN-GP framework

The term (b) is then incorporated in the SRM fitness function (Kecman 2001) to evaluate the performance of models during the evolutionary stages. The step-by-step procedure of incorporating complexity of model in the fitness function of GP is shown in Fig. 4. Modified SRM function is given by:

$$\text{SRM} = \frac{\text{SSE}}{N}\left(1 - \sqrt{\left(\left(\left(\frac{b}{N} - \left(\frac{b}{N}\log\left(\frac{b}{N}\right)\right)\right) + \left(\frac{\log\left(\frac{b}{N}\right)}{2N}\right)\right)\right)}\right)^{-1} \quad (1)$$

where b is degree of the polynomial that best fits the model during the evolutionary stages of GP, SSE is the sum of square of error of the generated model on the training data, and N is the number of training samples.

If any individual of the population does not satisfy the termination criterion, genetic operations (selection, subtree crossover, and subtree mutation) are implemented on the individuals to evolve the new population.

3. *Termination criterion and model selection*

The termination criterion is the maximum number of generations or the threshold error of the model (whichever is achieved earlier) as specified by the user. If this termination criterion is specified, then the final model is selected based on the minimum training error from all the runs.

M-file coded in MATLAB R2010b for CN-GP framework is shown in Fig. 5. The parameter settings of both methods (GP and CN-GP) are kept the same and adjusted using a trial-and-error approach (Fig. 6) based on the study conducted on applications of evolutionary algorithms in modeling of the machining processes (Garg et al. 2015b; Garg and Lam 2015b). The two methods are applied

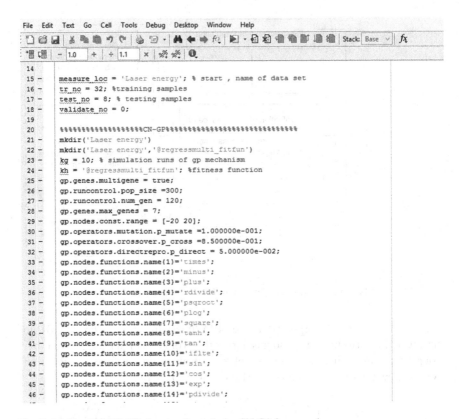

Fig. 5 M-file of MATLAB showing the code for CN-GP framework

```
Run parameters
---------------
Population size:           300
Number of generations:     120
Tournament size:           2
Lexicographic selection:   False
Max tree depth:            7
Max nodes per tree:        Inf
Using function set:        TIMES MINUS PLUS PLOG TANH TAN SIN COS EXP
Number of inputs:          2
Max genes:                 7|
Constants range:           [-20  20]
Using fitness function:    regressmulti_fitfun.m


Generation 0
Best fitness:    1.4406e-009
Mean fitness:    2.9074
Best nodecount:  82
RMSE fitness:    1.4406e-009
MAPE fitness:    1.2e-008
```

Fig. 6 Parameter settings for CN-GP framework

on the data set (data are discussed in Sect. 2) in the prediction of TAS and laser energy consumption of the SLS process. The four models (two GP and two CN-GP, (7)–(10) in Appendix) are selected. Performance and statistical comparison of these models is discussed in Sect. 4.

4 Results and Discussion

4.1 Statistical Validation of Energy Consumption Models Against the Experimental Data

The performance of the two methods (GP and CN-GP) is evaluated on the training and testing data (Figs. 7 and 8) in prediction of TAS and laser energy consumption of the SLS processes. Predictions obtained are compared to those of the actual data obtained from Nancharaiah et al. (2013). Five statistical metrics (the coefficient of determination (R^2), the mean absolute percentage error (MAPE), RMSE, the relative error (%), and the multiobjective error (MO)) are chosen to determine the method that gives the best generalization ability. The mathematical representation of these metrics is shown below:

$$R^2 = \left(\frac{\sum_{i=1}^{n} (A_i - \overline{A_i})(M_i - \overline{M_i})}{\sqrt{\sum_{i=1}^{n} (A_i - \overline{A_i})^2 \sum_{i=1}^{n} (M_i - \overline{M_i})^2}} \right)^2 \tag{2}$$

$$\text{MAPE} (\%) = \frac{1}{n} \sum_i \left| \frac{A_i - M_i}{A_i} \right| \tag{3}$$

$$\text{RMSE} = \sqrt{\frac{\sum_{i=1}^{N} M_i - A_i}{N}} \tag{4}$$

$$\text{Relative error} (\%) = \frac{|M_i - A_i|}{A_i} \times 100 \tag{5}$$

$$\text{Multiobjective error} = \frac{\text{MAPE} + \text{RMSE}}{R^2} \tag{6}$$

where M_i and A_i are the predicted and actual values, respectively, $\overline{M_i}$ and $\overline{A_i}$ are the averages of the predicted and actual values, respectively, and n is the number of training samples.

On the training phase (Figs. 7a, c and 8a, c), both methods are able to "learn" from the data samples very well so as to form models with high correlation coefficients and lower levels of error. In the testing phase (Figs. 7b, d and 8b, d), the

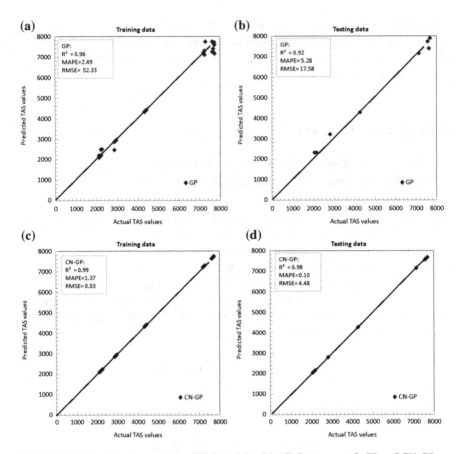

Fig. 7 Graphs showing error metrics of TAS models of the SLS process. **a, b** GP. **c, d** CN-GP

models formed from the new proposed CN-GP approach performed better than those of the standardized GP method in prediction of TAS and laser energy consumption. Lower values of MAPE and the high coefficients of determination (0.98 and 0.99) of the proposed model suggest that the predictions obtained are well in agreement with the experimental data obtained from Nancharaiah et al. (2013).

MO values for models formulated from both methods are computed based on (6) and shown in Table 3. Descriptive statistics for the models are shown in Table 4. It is obvious from Tables 3 and 4 that the lower values of MO and the confidence intervals range suggest that the models formulated from the CN-GP framework performed better than the standardized GP. For testing the goodness of fitness test for both the methods, t-tests and f-tests were conducted, and it was found that the p-values computed from models formulated from both methods is > 0.05, (Table 5) which indicates that there is hardly any difference between actual and predicted values. Based on the statistical evaluation of the models, it

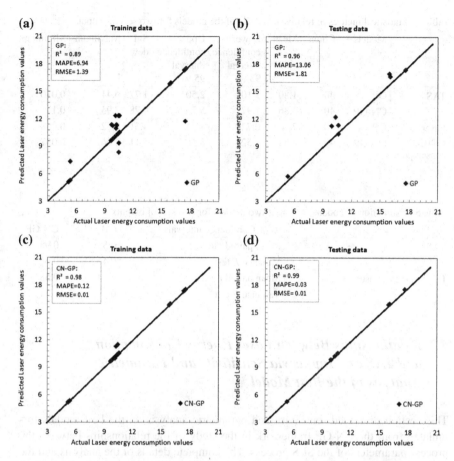

Fig. 8 Graphs showing error metrics of laser energy consumption models of the SLS process. **a, b** GP. **c, d** CN-GP

Table 3 Multi-objective error of the models for two sets of outputs

	Models	Training phase	Testing phase
TAS	GP	57.1	25
	CN-GP	1.42	4.67
Laser energy consumption	GP	9.35	15.48
	CN-GP	0.13	0.04

can be concluded that the models formed from the proposed CN-GP framework have outperformed the standardized GP and are able to predict the TAS and laser energy consumption satisfactorily. The performances of the two methods were somewhat similar in the case of laser energy consumption.

Table 4 Statistical metrics of relative error (%) of the models for two sets of outputs

	Models	Count	Mean	Lower confidence interval 95 %	Upper confidence interval 95 %	Std dev	Maximum	Minimum
TAS	GP	40	1.80	1.10	2.50	1.77	6.41	0.010
	CN-GP	40	3.86	1.97	5.76	4.78	7.95	0.11
Laser energy consumption	GP	40	2.50	0.72	4.29	4.51	8.62	0.038
	CN-GP	40	3.54	1.52	5.56	5.11	4.06	0.057

Table 5 Estimation of goodness of fit of two models for two sets of outputs

	95 % Confidence intervals	GP	CN-GP
TAS	Mean paired t test	0.53	0.98
	Variance F test	0.46	0.62
Laser energy consumption	Mean paired t test	0.34	0.58
	Variance F test	0.29	0.41

4.2 Relationships Between Laser Energy Consumption and TAS and Inputs via Sensitivity and Parametric Analysis of the Best Model

This section discussed sensitivity and parametric analysis on the best models formulated from the CN-GP framework in the study of the relationships between the process parameters of the SLS process. The complete details of the analysis and the mathematical formulae used are given in the study conducted by Garg et al. (2014c).

The results of the sensitivity analysis is shown in Table 6 and it is clear that the input parameter slice thickness has the highest impact on the TAS and laser energy consumption whereas the input part orientation has almost no effect on the two outputs (TAS and laser energy consumption). This means for proper monitoring of the energy consumption in the SLS process, the parameter slice thickness needs to be adjusted appropriately. The results of the parametric are shown in Fig. 9. Parametric analysis is performed by varying each input process parameters

Table 6 Contribution of input parameters to the two outputs (laser energy consumption and TAS)

	Input variable	Contribution (%) to outputs
TAS	Slice thickness	97.1
	Built part orientation	2.9
Laser energy consumption	Slice thickness	97.4
	Built part orientation	1.6

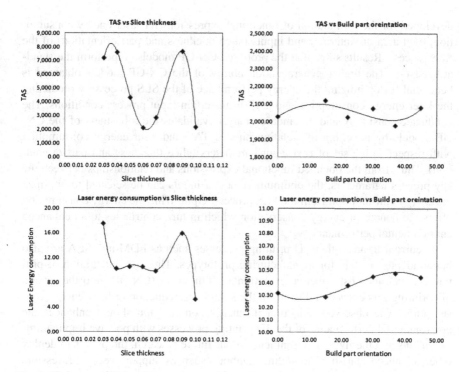

Fig. 9 Variation of TAS and laser energy consumption with respect to input parameters of SLS process

in succession and keeping the others fixed at their mean values. Figure 9 shows the plots generated for the two outputs (TAS and laser energy consumption) based on each of the input variables. It can be seen that the TAS and laser energy consumption behave nonlinearly (first and decreasing then increasing) with respect to the slice thickness, whereas it remains almost constant with respect to the built part orientation. The analysis is in good agreement with that of the study conducted by Nancharaiah et al. (2013). Thus, from the sensitivity and parametric analysis, we can select the optimal values of slice thickness which minimize the TAS and thus the laser energy consumption. In this way, the CN-GP models can be used to study the relationships between the process parameters and for monitoring the energy consumption in the SLS process.

5 Conclusions

The present work contributes to research and practice by the formulation of explicit and generalized laser energy consumption mathematical models for 3D printing processes. In this context, a complexity-based evolutionary framework (CN-GP) is developed for the formulation of functional expressions for laser energy consumption, total area of sintering, and inputs (slice thickness and part orientation) of the SLS process. Results show that the proposed CN-GP models outperform the standardized GP. The higher generalization ability of the CN-GP models obtained is beneficial for optimizing the energy performance of the SLS process by predicting the laser energy consumption and TAS in uncertain input process conditions. The conducted sensitivity and parametric analysis validates the robustness of the CN-GP models by unveiling the relationships of TAS and laser energy consumption with respect to the set of two input parameters (slice thickness and part orientation). Thus, from the obtained functional expressions and relationships between the key process parameters, the optimum values of inputs can be selected to optimize the energy consumption from the SLS process. Optimizing the energy consumption offers the benefit of energy conservation which in turn contributes to an enhanced environmental performance.

In current trends, other 3D printing processes such as FDM and SLA are also being adapted widely for manufacturing prototypes. Thus, the integration of proposed novel and cost effective evolutionary framework (CN-GP) into the several 3D printing processes could result in the significant reduction of laser energy consumption. The observed reduction in energy consumption should enhances the environmental performance of the 3D printing processes with positive implications to humans and the living environment. In our future research, the plan is to deploy other advanced statistical modeling methods such as support vector regression, response surface methodology, and dynamic neural networks and to evaluate any economic and environmental differences with respect to the current study.

Acknowledgments This study was supported by Nanyang Technological University's funding, reference number M060030008.

Appendix

$$
\begin{aligned}
\mathbf{TAS_{GP}} = {}& -88628431.5699 + (-14421262.2035) \\
& \times (\tanh(\sin(\tan(\cos((x1) \times ((-8.993442))))))) + (107.691) \\
& \times (\sin(x2)) + (70467967.5522) \times (\sin((x1) \times (\cos((\tanh(x1)) \\
& * ((-8.993442)))))) + (-3390.7491) \times (\cos((\cos((x1) \\
& \times ((-8.993442)))) \times ((-8.993442)))) + (-65390146.1759) \\
& \times (\sin(\tanh(\tanh(x1)))) + (-24.1412) \times (\tan(x2)) \\
& + (99531980.4057) \times (\cos(\tanh(\tan(x1))))
\end{aligned}
$$

(7)

$$
\begin{aligned}
\text{TAS}_{\text{CN-GP}} = {} & 5137644.8262 + (-1706557.4605) \times (\tan((\sin((\cos(x1)) \\
& + (((9.289170)) \times (x1)))) - (\tan(\tan(x1))))) \\
& + (-3439756477.4271) \times (\tan((x1) \times (x1))) + (-107.6723) \times ((x1) \\
& - (\sin(x2))) + (-26870.1687) \times ((\tan(x1)) \times (\tan(\tan(((9.289170)) \\
& \times (x1))))) + (24.1369) \times (((\sin((\cos(x1)) + (((9.289170)) \\
& \times (x1)))) - (\tan(x2))) - (\tan(x1))) \\
& + (4643987.9876) \times ((\sin((\cos(x1)) + (((9.289170)) \times (x1)))) \\
& - (\tan(\cos(\tan(x1))))) + (-574696436.1438) \\
& \times ((x1) \times ((x1) \times (((-5.810103)) - (x1))))
\end{aligned}
$$

$$\text{(8)}$$

$$
\begin{aligned}
\text{Laser energy consumption}_{\text{GP}} = {} & 1115180.5442 + (744275.6671) \times (\tan(\sin((x1) \\
& + (x1)))) + (-357003.1223) \\
& \times (\tan(\tanh(\tan(((-8.767966)) \times (x1))))) \\
& + (347109.7909) \times ((\tan(\tanh(\tan(((-8.767966)) \\
& \times (x1))))) - (((-4.185710)) + (\sin((x1) \\
& + (x1))))) + (-4442389.3594) \\
& \times (\cos(((x1) - ((0.17))) \times (((4.09)) \times (x1)))) \\
& + (-4151) \times (\cos(\sin(((9.961556)) + (x2)))) \\
& + (2149599.4264) \times (\cos(\sin(((9.961556)) \\
& + (x1)))) + (0.045548) \times ((((\tan(\tan(x1))) \\
& \times (\cos(\sin(x1)))) + (\sin(x1))) + (\sin(((x1) \\
& - (\sin(x2))) + (\sin((x1) + (x1)))))))
\end{aligned}
$$

$$\text{(9)}$$

$$
\begin{aligned}
\text{Laser energy consumption}_{\text{CN-GP}} = {} & -594678.7278 + (2241.127) \times ((\cos(\tan(((x1) \\
& - ((-2.441394))) \\
& - ((x1) \times ((5.972072)))))) \\
& \times ((\tanh(\tan(\sin(x1)))) - (\tanh(\tanh(\cos(x1))))) \\
& + (-3719342.8431) \times ((x1) \times (x1)) \\
& + (-177698.568) \times (\tan(\tan(\sin((((5.372252)) \\
& + ((-0.213505))) + (\tanh(x1)))))) \\
& + (2474839.7138) \times (\tan(\sin((\cos(x1)) \times (x1)))) \\
& + (-0.064203) \times ((\sin(x2)) + (\tanh(((\cos(x2)) \\
& + (\cos(x1))) - ((\tan(x2)) - ((x1) - (x1)))))) \\
& + (0.085606) \times (\tan((\tanh(((6.960941)) \\
& \times (\sin(x2)))) \times (\sin((\sin(x2)) + ((x2) + (x1)))))) \\
& + (41104.3208) \times (\cos(((((5.372252)) + (x1)) \\
& + (\tanh(x1))) \times ((5.972072))))
\end{aligned}
$$

$$(10)$$

References

Baumers M, Tuck C, Bourell DL, Sreenivasan R, Hague R (2011) Sustainability of additive man-ufacturing: measuring the energy consumption of the laser sintering process. Proc Inst Mech Eng Part B J Eng Manuf 225(December (B12)):2228–2239

Beal V et al (2009) Statistical evaluation of laser energy density effect on mechanical proper-ties of polyamide components manufactured by selective laser sintering. J Appl Polym Sci 113(5):2910–2919

Bellini A, Guceri S, Bertoldi M (2004) Liquefier dynamics in fused deposition. J Manuf Sci Eng Trans ASME 126(2):237–46

Cervera GBM, Lombera G (1999) Numerical prediction of temperature and density distributions in selective laser sintering processes. Rapid Prototyping J 5(1):21–26

Chatterjee A et al (2003) An experimental design approach to selective laser sintering of low car-bon steel. J Mater Process Technol 136(1):151–157

Childs THC, Tontowi AE (2001) Selective laser sintering of a crystalline and a glass-filled crys-talline polymer: experiments and simulations. Proc Inst Mech Eng Part B J Eng Manuf 215(11):1481–95

Childs THC, Berzins M, Ryder GR, Tontowi A (1999) Selective laser sintering of an amor-phous polymer—simulations and experiments. Proc Inst Mech Eng Part B J Eng Manuf 213(4):333–49

Cho HS, Park WS, Choi BW, Leu MC (2000) Determining optimal parameters for stereolithogra-phy processes via genetic algorithm. J Manuf Syst 19(1):18–27

Cong-Zhong C et al (2009) Density prediction of selective laser sintering components based on support vector regression. Acta Physica Sinica 58(6):S8–S14

Deckard CR, McClure P (1988) Selective laser sintering

Fuh JYH, Choo YS, Nee AYC, Lu L, Lee KC (1995) Improvement of the UV curing process for the laser lithography technique. Mater Des 16(1):23–32

Garg A, Garg A, Tai K, Sreedeep S (2014a) An integrated SRM-multi-gene genetic programming approach for prediction of factor of safety of 3-D soil nailed slopes. Eng Appl Artif Intell 30:30–40

Garg A, Tai K, Savalani MM (2014b) State-of-the-art in empirical modeling of rapid prototyping processes. Rapid Prototyping J 20(2):164–178

Garg A, Tai K, Savalani MM (2014c) Formulation of bead width model of an SLM prototype using modified multi-gene genetic programming approach. Int J Adv Manuf Technol 73(1-4):375–388

Garg A, Lam JSL (2015a) Improving environmental sustainability by formulation of generalized power consumption models using an ensemble evolutionary approach. J Cleaner Prod 102:246–263

Garg A, Lam JSL (2015b) Measurement of environmental aspect of 3-D printing process using soft computing methods. Measurement 75:210–217

Garg A, Lam JSL, Gao L (2015a) Energy conservation in manufacturing operations: modelling the milling process by a new complexity-based evolutionary approach. J Cleaner Prod. doi:10.1016/j.jclepro.2015.06.043

Garg A, Vijayaraghavan V, Lam JSL, Singru MP, Gao Liang (2015b) A molecular simulation based computational intelligence study of a nano-machining process with implications on its environmental performance. Swarm Evol Comput 21:54–63

Garg A, Lam JSL, Savalani MM (2015c) A new computational intelligence approach in formulation of functional relationship of open porosity of the additive manufacturing process. Int J Adv Manuf Technol 80(1):555–565

Kecman V (2001) Learning and soft computing: support vector machines, neural networks, and fuzzy logic models. MIT press

Kellens K, Yasa E, Dewulf W, Duflou J (2010) Environmental assessment of selective laser melting and selective laser sintering. In: Going green—CARE INNOVATION 2010

Koza JR (1994) Genetic programming II: automatic discovery of reusable programs. MIT, USA

Kruth JP, Kumar S (2005) Statistical analysis of experimental parameters in selective laser sintering. Adv Eng Mater 7(8):750–755

Lam JSL (2015) Designing a sustainable maritime supply chain: a hybrid QFD-ANP Approach. Transp Res Part E 78:70–81

Lam JSL, Dai J (2015) Environmental sustainability of logistics service provider: an ANP-QFD approach. Int J Logistics Manage 26(2):313–333

Lam JSL, Lai KH (2015) Developing Environmental Sustainability by ANP-QFD Approach: the Case of Shipping Operations. J Clean Prod 105:275–284

Liao H-T, Shie J-R (2007) Optimization on selective laser sintering of metallic powder via design of experiments method. Rapid Prototyping J 13(3):156–162

Luo YC, Ji ZM, Leu MC, Caudill R (1999a) Environmental performance analysis of solid freeform fabrication processes. In: Proceedings of the 1999 IEEE inter-national symposium on electronics and the environment, ISEE—1999, p. 1–6

Luo YC, Leu MC, Ji ZM (1999b) Assessment of environmental performance of rapid prototyping and rapid tooling processes. In: Solid freeform fabrication pro-ceedings, Aug 1999, pp 783–91

Mognol P, Lepicart D, Perry N (2006) Rapid prototyping: energy and environment in the spotlight. Rapid Prototyping J 12(1):26–34

Nancharaiah T, Nagabhushanam M, Amar Nagendram B (2013) Process parameters optimization in SLS process using design of experiments. Int J Mech Eng Technol 4(2):162–171

Nelson JC et al (1993) Model of the selective laser sintering of bisphenol-A polycarbonate. Ind Eng Chem Res 32(10):2305–2317

Niino T, Haraguchi H, Itagaki Y (2011) Feasibility study on plastic laser sintering without powder bed preheating. In: Proceedings of the 22th solid freeform fabrication symposium

Paul Ratnadeep, Anand Sam (2012) Process energy analysis and optimization in selective laser sintering. J Manuf Syst 31(4):429–437

Raghunath N, Pandey PM (2007) Improving accuracy through shrinkage modelling by using Taguchi method in selective laser sintering. Int J Mach Tools Manuf 47(6):985–995

Savalani MM et al (2012) The effects and interactions of fabrication parameters on the properties of selective laser sintered hydroxyapatite polyamide composite biomaterials. Rapid Prototyping J 18(1):16–27

Shen, X et al (2004) Density prediction of selective laser sintering components based on artificial neural network. In: Advances in neural networks-ISNN 2004. Springer, Berlin, pp 832–840

Singh A, Prakash RS (2010) DOE based three-dimensional finite element analysis for predicting density of a laser-sintered component. Rapid Prototyping J 16(6):460–467

Sreenivasan R, Bourell DL (2009) Sustainability study in selective laser sintering—an energy perspective. In: Proceedings of the 20th solid freeform fabrication symposium, pp 3–5

Sun MM (1991) Physical modeling of the selective laser sintering process. Dissertation. TX, United States: The University of Texas at Austin

Sun MM, Beaman J (1991) A three dimensional model for selective laser sintering. In: Solid freeform fabrication symposium

Sun MM, Beaman J, Barlow JW (1990) Parametric analysis of the selective laser sin-tering process. In: Solid freeform fabrication symposium

Thompson DC, Crawford RH (1997) Computational quality measures for evaluation of part orientation in freeform fabrication. J Manuf Syst 16(4):273–289

Tontowi AE, Childs T (2001) Density prediction of crystalline polymer sintered components at various powder bed temperatures. Rapid Prototyping J 7(3):180–184

Vijayaraghavan V, Garg A, Wong CH, Tai K, Regella PS, Tsai CM (2014) Density characteristics of laser-sintered three-dimensional printing parts investigated by using an integrated finite element analysis–based evolutionary algorithm approach. Proc Inst Mech Eng Part B J Eng Manuf (Imeche), Dec 2, 2014. 0954405414558131

Williams JD, Deckard CR (1998) Advances in modeling the effects of selected parameters on the SLS process. Rapid Prototyping J 4(2):90–100

Xu F, Wong YS, Loh HT (2000) Toward generic models for comparative evaluation and process selection in rapid prototyping and manufacturing. J Manuf Syst 19(5):283–296

Yang H-J, Hwang P-J, Lee S-H (2002) A study on shrinkage compensation of the SLS process by using the Taguchi method. Int J Mach Tools Manuf 42:1203–1212

Yardimci AM, Guceri SI, Danforth SC, Agarwala M (1995) A phenomenological numerical model for fused deposition processing of particle filled parts. In: Proceedings of the solid freeform fabrication symposium. The University of Texas, pp 189–195

3D Printing Sociocultural Sustainability

Jennifer Loy, Samuel Canning and Natalie Haskell

Abstract Additive manufacturing, more commonly termed 3D printing, could be criticized as contrary to the principles of sustainability, as it enables unregulated production that can have a negative impact on the environment if misused. However, this technology can also support value added, invested design by putting accessible digital fabrication within the reach of the designer-maker. In an increasingly homogenized world, this technology has the potential to improve sociocultural sustainability (retaining social and cultural diversity as a factor of sustainability) by supporting the practice of the individual designer-maker. The technology has the potential to contribute to the economic viability of the designer-maker, providing an effective mechanism for an artisan to compete in a global market utilising distributed manufacturing, the availability of increased geometric complexity, and the ability to customize products. This chapter provides an argument for the potential role of 3D printing in supporting sociocultural sustainability and is based on practice-led research. The impact of digital fabrication on practice for designer-makers is explored in relation to its potential to support the retention of craftsmanship skills, values and cultural referencing particular to a community, and community of craft practice.

Keywords Bespoke · Computer numerically controlled routering · Crafts · Design activism · Distributed manufacturing · Invested design · Values

J. Loy (✉) · S. Canning · N. Haskell
QCA and Griffith School of Engineering, Griffith University, Southport, Gold Coast, QLD 4222, Australia
e-mail: j.loy@griffith.edu.au

© Springer Science+Business Media Singapore 2016
S.S. Muthu and M.M. Savalani (eds.), *Handbook of Sustainability in Additive Manufacturing*, Environmental Footprints and Eco-design of Products and Processes, DOI 10.1007/978-981-10-0549-7_4

1 Introduction

Product design definitions of sustainability have developed over the last 20 years
to encompass an understanding of production systems which track and evaluate
the environmental impact of even the smallest manufactured items. In addition,
with the widespread adoption of triple-bottom-line accounting as discussed in
Cannibals with Forks (Elkington 1999), definitions of sustainability now encompass wide-ranging considerations beyond the environmental impact of products.
They include the implications of choices, made as part of broader patterns of
production and consumption, on the social, political and economic welfare of all
stakeholders in the full life cycle of a product. In the preface to the book, *Fashion
& Sustainability: Design for Change* (Fletcher and Grose 2012), Hawken states
that sustainability offers all design an opportunity for critique, in that it challenges
decision-making at the micro level of detail as well as at a whole systems level.
This includes the economic systems supported by design decisions, the values
modelled, and the belief systems that are fostered. Taking this all into account is
a complex proposition, although Fletcher and Grose (2012) argue that the range of
skills needed to practice with a sustainable design framework are within the scope
of designers as they are already comfortable facing the unknown, working across
disciplines and, in particular, synthesising complex information. They suggest that
designers are equipped to deal with the challenges presented by sustainability drivers and the "intuitive leaps in thinking" (Fletcher and Grose 2012, p. 172) required
in order to rethink systems and create step-changes to respond to the evolving
nature of sustainability research at a particular point in time.

To practice the 'leaps in thinking' currently required, the designer needs to have
a clear understanding of the current context. Makepeace (1995) described good
design as informed by the knowledge available at a point in time and the aspirations for society, to express values and ideals relevant for that period. Definitions
of sustainability currently informing designers are not based solely on a data-led
approach to life cycle inventory and assessment, but on broader impact markers.
Social sustainability is integral to sustainability discourse and practice. It requires
a qualitative rather than quantitative perspective on the impact of patterns of production and consumption, and the relationship of people to products. This human
centred approach has long been fundamental to the design process in relation to
the experience of product users (Norman 2013). The approach was then extended
to the experience of workers during the 1990s, with designers considering the
impact of their design decisions and the organization of production on the experiences and output of workers. However, the current complexities of design for sustainability now encompass care for the very fabric of a society itself. This takes
design responsibility to another level.

Fuad-Luke, in his book *Design Activism: Beautiful Strangeness for a
Sustainable World* considers that design: "embraces myths and meaning, philosophy, science, teaching/education, anthropology, sociology, material culture
studies, media and cultural studies, economics, political sciences, economics and

ecology", arguing that it is "design's ability to operate through 'things' and 'systems' that makes it particularly suitable for dealing with contemporary societal, economic and environmental issues" (Fuad-Luke 2009, p. 2). This evolution has changed the very nature of design practice for the professional designer. From its practical, industrial roots as a documenter with the ability to translate fashionable trends into commercial forms, the role of the designer now includes responding to complex social and environmental issues. The design of products is therefore arguably now less about product outcomes than it is about negotiating human relationships (with each other, environments and economies). Designers, by current definition, cannot avoid being involved in questions about what sort of society their work supports.

Fuad-Luke argues that, as a result, all design is political and therefore all designers are involved in a form of design activism. He suggests that: "Forms of activism are also an attempt to disrupt existing paradigms of shared meaning, values and purpose to replace them with new ones, and so activism perhaps embodies a sense of developing the spiritual capacity of individual human capital that is collectivized in social capital" (Fuad-Luke 2009, p. 10). He directly relates this back to sustainability aspirations: "Recent interpretations of human capital have included physical, intellectual, emotional and spiritual capacities...the notion of 'spiritual capital' goes beyond any religious or institutionalized vision of 'spirituality', rather it addresses fundamental searches for 'shared meanings, value and ultimate purpose' and this is critical if we are to achieve 'sustainable capitalism and a sustainable society" (Fuad-Luke 2009, p. 10).

The other major influence on design today is the growth in the digital landscape. The Internet has changed communication practices and radically impacted business practices, with new relationships emerging across the globe between makers and users. Changing relationships and systems made possible through the Internet are transforming the possibilities for small-scale producers, such as designer-makers. This is in part through the direct communication channels now possible between makers and users, opening up new markets, but also through the development of mechanisms that enhance the information flow about their work that supports the use of designer-makers. Alongside digital communication tools, the other most significant factor for designer-makers in this digital landscape has been the development of direct additive manufacturing technologies. These technologies, more commonly known as 3D printing, and the communication systems growing alongside them, are providing new fabrication opportunities that have the potential to change production models and consumption behaviours and attitudes. Understanding their potential role in supporting a sustainable society as defined by Fuad-Luke, requires looking at the technologies in relation to traditional practices.

With its technological basis, it may seem a contradiction to consider additive manufacturing in terms of craft practices, yet, over the last 20 years, computer-based, subtractive digital technologies, such as computer numerically controlled (CNC) routering and laser cutting, have been clearly assimilated into craft practice and the production capacity of designer-makers. These technologies have added to the commercial viability of practice, as demonstrated later in this chapter in the work of David Trubridge, and have opened up new opportunities for designing and making not previously possible. The emergence of direct additive manufacturing in accessible formats has led to an initial burgeoning of low-end products with little professional design input, readily available online. As with many new technologies, following an initial over exuberance in the use of the tools available, not dissimilar to the overuse of mechanically produced decoration on products shown at the Great Exhibition in 1851, gradually a combination of thought and quality of design that maximizes the potential of the technology reasserts the underlying ideas and drivers of artisanship. Evidence of this practice for 3D printing is now emerging, such as demonstrated later in the chapter in the work of designer-maker David Haggerty.

The integration of any new technologies into craft-based practices can be seen as part of an ongoing dialogue which transcends the inevitable changes in technology. The UK Crafts Council *Making Value* report states that "In craft discourse, craft is increasingly understood as a distinctive body of knowledge, skills and aptitudes, centred around a process of reflective engagement with the material and the digital worlds" (Swartz and Yair 2010, p. 110). According to *Digital Crafts* author, Shillito (2013), computer aided design and fabrication allows for designer-makers to increase capacity, thereby extending their influence and supporting their commercial viability, whilst also enabling higher levels of experimentation. This chapter includes arguments for the importance of this capability in relation to maintaining sociocultural sustainability, that is, in retaining and fostering social and cultural diversity as a factor in a definition of sustainability based on triple-bottom-line accounting. Supporting the work of individual designer-makers and of craft ideals provides a means to challenge the current dominance of mass production and the inherent issues. Mass production system strategies can contribute to a built-in obsolescence in the design of commercial products that is at odds with the sustainability paradigm. This is not only in terms of the environmental impacts of those products, but also in relation to the impact they can have on cultural diversity and richness. This chapter discusses the potential of 3D printing to contribute to the ongoing crafts discourse in connecting people to places and the environment through the development of invested products printed on demand.

2 Creating Connections

Re-evaluating the role of the designer to respond to a view of sustainability that includes supporting social capital involves considering the idea of a designer as a design anthropologist (Gunn et al. 2013) engaging with cultural studies in order to understand social constructs. Cultural identity forms part of the definition and expression of shared meanings, values and purpose for major and minor societal groups. By re-enforcing cultural identity for subcultures, as well as the dominant culture in a society, sustainable design can protect diversity. This is becoming a particularly pertinent issue with the rise of global connectivity and a corresponding convergence of consumer systems that in turn raises questions about the impact of low cost, homogenous quality products on "social and environmental richness" (Fletcher and Grose 2012, p. 128). Fletcher and Grose argue that designers responding to the sustainability imperative need to challenge the dominant mass production practices undermining this richness. The idea is that designers need to work to serve the well-being of citizens as well as the environment, and that they can do this by contributing to a redirection of the mechanisms and policies that shape the "cultural logic of society" (Fletcher and Grose 2012, p. 173).

According to Kawamura (2012), the study of cultures involves understanding the "way of life" and "maps of meaning" (Kawamura 2012, p. 7) created by groups in order to make sense of the world for the individuals within those groups. Building on arguments to support diversity and social and environmental richness and challenge dominant production systems and relationships, designers need to support subcultures that challenge the behaviours, values and meanings inherent in more dominant cultural practices that support mass production systems. Studies of subcultures highlight notions of resistance to a more dominant culture as a primary driver: "The prefix sub implies notions of distinctiveness and difference from the dominant or mainstream society. Therefore, a subculture is constituted by groups of individuals who share distinct values and norms that are against dominant or mainstream society. Members often create their own symbols...that are comprehensible only to the inside members. If you are not dominant, you are subordinate; if you are not in the mainstream, you are on the periphery. Subcultures have been seen as spaces for deviant communities to claim their position or space, metaphorically and literally, for themselves" (Kawamura 2012, p. 7).

Subcultures in this context include, for example, the 'Slow Food' movement, conceived by Carlo Petrini in 1986 as a response to the growing proliferation of fast food outlets. Petrini fought unsuccessfully against the opening of a McDonald's outlet in Rome's Piazza di Spagna (Hickman 2009) and went on to espouse the virtues of humanism and a new school of economic thought. Just as fast food is not merely a descriptor of speed, neither is 'Slow Food'. 'Slow' has become an expression of attitude and values about social and environmental richness and responsibility that has been adopted by a cross-section of society intersecting a myriad of production and design situations. The 'Slow' movement aligns with the work of activists in the 'Transition Towns' initiative with the goal to build connections between people, and also between people and their environment as part of a response to climate change concerns and a demise of localized, values based systems. In both, the goal is a re-education on definitions of wealth based on community and environmental reciprocity as the building block for changing economic relationships worldwide: "Systemic innovation around sustainability begins with a change of thought pattern and behaviours, which leads to the building of structures and practices defining and describing economic activity by ecological limits" (Fletcher and Grose 2012, p. 174).

For product designers, working with notions of supporting connections between people, subcultures and the environment involves heightening relationships between people and the products that populate their home, workplace and community. In *Sustainable by Design*, Walker (2006) argues for the development of invested products with connections to people and places in order to promote longevity, where people retain products for longer periods, are more inclined to repair them as needed, and spread the embodied energy required to make them over a longer period of time. He reasons that this is in direct conflict with the dominant production culture: "Consumer capitalism actually abhors products that possess enduring usefulness and value" (Walker 2006, p. 170). Walker also argues that distancing the consumer from manufacturing, as is current practice in mass production systems, reduces the sense of connection to the object for the user, and thereby disconnects the user from a sense of responsibility for that object. This distancing is detrimental in terms of valuing the resources invested in the production of the object, reducing the environmental impact of the use phase of the product where possible and consideration of the end-of-life of a product and its potential to be adapted for reuse or the materials to be recycled. Walker also argues this disconnect affects the sense of responsibility and empowerment of the designer and client: "It is all too easy to distance ourselves from the effects of our decisions and our actions when design and manufacturing are carried out within a globalized system. When head offices are located in North American or European cities but resources are extracted, processed, formed and assembled 'somewhere else', decision makers inevitably lack a full appreciation of the consequences of their actions. Within this system, information becomes filtered down to the essential but abstracted data of production performance, unit costs and profits. A more holistic understanding of the meaning of decisions and their impacts is lacking because there is little direct connection with people and place" (Walker 2006, p. 167). By focussing on supporting subcultures, and reconnecting people and places, the designer is making a case for demonstrating clear 'cause and effect' in the choices made by producers and consumers. Products made within this strategy are, by definition, bespoke, re-enforce cultural relationships, value resources and are valued by the community for which they are designed.

3 Craft Discourse and Sustainability

If an evolving approach to sustainability includes sociocultural sustainability—that is, it protects social 'richness', diversity, and connection to places, people and materials in a measured, considered way that produces invested products—then this builds on the discourse of the relationship between craft principles and mass production that has been ongoing since the Great Exhibition in 1851. The arguments on the tension between the work of craftsmen, focusing on one-off pieces or batch produced products, and the mechanisms and output of mass manufacturing, have been inseparable from discussion on social reform, quality of life and environmental impact. William Morris and the Arts and Crafts movement began the discourse based on the work of Ruskin, and the issues have remained intertwined as the division of labor has increased, production systems and supply chains have become more disparate and economic and commercial drivers to maintain failing mass production systems, such as can be seen in the aftermath of the car industry in Detroit, have overtaken the drive to meet market demands. The sustainability imperative has the potential to empower less economically competitive mechanisms of production at this time, as it rebalances the financial evaluation of output. A fuller accounting of the triple-bottom-line for production systems, driven by tougher policies and regulations in response to climate change evidence, could further serve to tip the balance in favour of sociocultural sustainability. If this is the case, then a revisiting of the relationship between craft principles, design and consumerism is due.

Since Rachel Carson wrote the seminal book *Silent Spring* in 1962, there has been a focus on research into environmental issues which has led to a growing body of knowledge impacting decision-making at the highest levels of governments. This has also acted as a driver for change at a community level. However, it is not the only revolution to dominate thinking during this time that is now impacting the discourse around production and consumption. Since the mid-1950s, the development of digital communication and production tools, and the resulting 'digital revolution', have added to the roller coaster of changes in thinking about making and consumerism. The main two factors have been Web 2.0, which has enabled interaction between all stakeholders involved in the creation, use and end-of-life of a product, and digital fabrication. Digital fabrication has grown alongside computing, with the integration of CNC routers into the production environment, and laser cutting technologies of multiple materials. Most recently, additive manufacturing developments have added to the digital fabrication capabilities of manufacturers. These technologies were initially designed for prototyping in the mid-1980s, but have been developed into end-use manufacturing technologies for finished products over the last 15 years. Additive manufacturing, now commonly known as 3D printing, is characterized as a way of making a product directly from a 3D computer model. 3D printing includes a number of diverse technologies, but all essentially allow for an individual product to be printed without the need for a mould, and without subtracting material during the build. There is now a considerable body of knowledge on the range technologies and their applications for different industries, in, for example, *Additive Manufacturing: 3D Printing, Rapid Prototyping and Direct Digital Manufacturing* (Gibson et al. 2015). In relation to the discourse around reconnecting users to products, fused deposition modelling is the most significant technology as it is directly accessible to the general public. However, the higher end technologies, such as selective laser sintering and direct laser melting, are also relevant because of the shift in business mechanisms emerging in tandem with the technology. It is this parallel development of digital communication and digital fabrication that provides new opportunities to converge on a new approach to sustainability and product design which responds to ideas of supporting social and environmental 'richness' and diversity, and connection to people and places, as exemplified in the 'Slow' movement.

The verb 'to craft', defined by the *Collins English Dictionary*, is to "make with skill". Skill requires a commitment of time to develop which aligns with the current debate around the idea of 'Slow' and related social movements. Since Morris and Ruskin, notions of craft have been embedded in social issues relating to the organization of labour, production practices and quality of life. Sustainability and design activism are linked to the attitudes and behaviours modelled by designer-makers and professional craftspeople in society. The designer-maker demonstrates connection to community, to materials, to place, to quality of life, and it is possible, therefore, to link the practice of the designer-maker to current thinking on sustainability, and in particular to retaining cultural values, identity, diversity and aesthetics.

However, the notion of craftsmanship has to be seen on a continuum. It is not a static notion as the discourse is ongoing as ideas of community, place, quality of life, connection to places and products, cultural values, etc., evolve. Change is the constant, and whilst underlying sustainability drivers are reasonably defined, the manifestations of how the response to ideas of sustainability are expressed is constantly changing, just as the ideas of art as social commentary are continuing, but the way in which that manifests into outcomes changes. There is always rigorous debate about what qualifies as expressing particular sentiments and ideas, but the reality is that the rules cannot remain static as the place in time, the context, is always changing. Ideas of sociocultural sustainability are embedded in a specific place in time and by that very definition are always changing.

In summary, issues of sustainability and sustainable design strategies cannot be considered without taking into account the parallel development of the digital revolution. Over the last 30 years, advances in digital technology have transformed the world. Most significantly have been the rise of Web 2.0 and enabled communication, the generation and manipulation of data and interactive electronics, and digital fabrication. The impact on designer-makers has been twofold. On the one hand, the technologies have changed the possibilities in the designer-makers' world of practice. On the other hand, it has changed the maker paradigm in which they operate and the attitude of the community they operate within.

4 Making Context

"Craftspeople who focus on the handmade for one-off pieces or limited series have tended to reject mass production. These makers resolve their ideas through working directly with materials, using related tools and technologies, seeking self-sufficiency and personal artistic expression through the contemporary exploration of a range of cultural traditions. Alongside the attraction of mass-manufactured 'design brands' there are also collectors and consumers who value objects that are made by hand and who enjoy their association with the person who made them and the evidence of how they may have been made" (Cochrane 2007, p. 9). If an approach to sociocultural sustainability includes supporting the development and creation of products embedded in the social fabric of a community and express the ideas and values of a subculture, then that arguably involves sustaining the practice of the community-based artisan. The skills and knowledge embedded in high-end craft practice have been built over generations and are infused with cultural associations and referencing. The difficulty of extending the influence of invested objects more broadly has been twofold. First, education on the values imbued by the objects, including, for example, the use of local materials reducing embodied energy, and second, the relative costs. International product designer, Marc Newson, works within commercial mass production systems but has a background in sculpture and continues to value his hands-on connection: "How any approach to designing and making is valued changes over time. There has always been a creative tension between the poles of art and design, hand and machine, one-off and mass-produced, personal and popular, skill and imagination...my work is about a direct link between my head and my hands" (Cochrane 2007, p. 9). This personalization of making disconnects the buyers, making some traditional craft practices open to the criticism of elitism and personal art. This is the opposite of the driver for the 'Slow' movement. The imperative is reconnecting all stakeholders, and supporting the financial viability of high-end craft practice to ensure exemplars for quality output are still possible.

There are broader issues with regards to not supporting the making skills in a community which have implications for governments in terms of cultural identity and economics. The shift of production to developing countries in recent decades has resulted in an increase in low cost, low quality products which have little invested value, short life span, cradle to grave planning, rather than cradle to cradle as discussed by McDonough and Braungart (2002), non-regionalized specification of materials, increased transport miles, etc. If this direction continues, the irresponsible use of resources and overall impact on the biosphere is likely to continue. Alongside this, and relevant to this chapter, there is also a growing awareness of cultural identity at Government level, and of the importance of reconnecting people, place and values for sustainability to have meaning. For example, in 2005, the Smith Institute produced a report assessing the relationship between societal values and economic development which drew parallels between the rise of manufacturing in China, the draining of traditional manufacturing in the United Kingdom and the breakdown of cultural identity in Britain: "Countries like China will dominate manufacturing, while those in the West will lose their old skills and industries and have to redefine themselves not only economically, but also culturally, where their identity had been associated with values of manufacturing aesthetics and working ideologies" (Cochrane 2007, p. 10).

There is a growing understanding that skills-based industries, and the role of the designer-maker, are important in supporting a healthy community psyche to foster connections between people, places and environment which support ideas of sustainability and sociocultural sustainability for a rich, diverse society. The viability of these industries needs to be built on economically sustainable realities so that they can survive. The JamFactory in Australia, for example, is committed to providing designer-makers with the means to sustain their practice, nurturing their skills and values as vital for a vibrant culture and healthy society (JamFactory 2013). It is part of a discourse about a continuum with a value beyond that of personal expression and should therefore be integral to production systems—particularly now with the current understanding of sustainability.

5 Digital Crafting

For all the ideals of bespoke products in a newly sustainable world, the reality is that artisanship, and the customization of invested products, still struggles to maintain a place in the current commercial environment. Just as the techniques and skills of traditional potteries and mills in the UK, which were commercially successful for the first 100 years of their existence, have been lost as the factories proved unable to adapt to changing international markets, so have growing economic pressures from overblown mass manufacturing of low margin products marginalized highly skilled, high cost craft pieces. This has shrunk the market for the work of professional craftspeople, even with the accessibility of crafters to wider markets through the web (the work is still generally localized in meaning and highly invested to high cost). This inevitably reduces their social impact and thereby their relevance in sociocultural sustainability: "If the applied arts are to mean anything, they desperately need to engage with the concept of modernity in a considered way" (Cochrane 2007, p. 10). Craft practitioners have no choice but to look to the future to maintain the techniques, skills and approaches of their work. In responding to the driver of being part of a genuine shift in thinking at a meta-level to support the social ideas behind movement such as Transition Towns and 'Slow' fashion, etc., there has to be a middle ground between economic viability and craft technologies. Working with value adding digital fabrication technologies, such 3D printing, digital communication technologies allow access to new markets and enhance communication for invested design with the potential to improve the commercial viability of individual makers. However, by fully embracing this approach, the designer-maker needs to train in digital as much as, if not more than, traditional making, particularly where 3D printing is concerned. Because of the requirements of highly skilled 3D computer-based modeling, with solid modeling software (rather than visualizing software) suitable for 3D printing to the extent that organic and complex modeling can be achieved, it takes practice and understanding. Taking the designer-maker out of the traditional workshop for a large proportion of time exposes the digital designer-maker to criticism: "Rapid prototyping takes place alongside hand forming, laser-technology alongside hand finishing; drawing beside digital design. There is, however, a real likelihood that skills and knowledge based on hands-on materials experience could be lost or difficult to retrieve...there are limitations to designs that take their form only on the screen; they demonstrate that their designs are more successful when they are integrated with ideas that grow out of a working knowledge of a materials-based designing and making process" (Cochrane 2007, p. 12). However, there is a limitation in this perspective on the design process which assumes an isolation in the use of design development tools which is not the reality of emerging practice. 3D computer modeling software began as a documentation tool, was incorporated at the end of design practice, and evolved to a point where it was technically possible to create objects solely in the virtual environment. However, just as good quality craft practice is based on an in-depth study and evaluation of materials, form, construction and aesthetics, so is good quality design that utilizes digital technologies still informed by in-depth research into these same factors. The argument is that: "The action of making, and the outcome of a crafted object, connects cultures, communities and generations. Handmade objects have a story. They have been touched, manipulated, hammered, thrown, blown, or carved by another human hand. They connect us to our past and to our familial and cultural histories" (Charny 2011, p. 7). However, this is based on a limited perception of making and expertise and skill. Digital skill redefines the concept of making and craft skill and it is arguably a cultural perception to suggest that it is not, rather than an actual reality. Whilst some may have difficulty with it, others embrace digital technology and integrate it in the same way that spindle moulders, band saws, and electric lathes are now integrated into making. Newson's work is informed by hand skills, and made with commercial production methods, providing a link between hand skills and mechanization. This suggests a continuum of craft practice, through the designer-maker to mechanizations, to include working with digital technologies such as 3D printing. David Pye presents a case that could be used against 3D printing, stating that craftsmanship equals risk, and that it is the inability to go back and redo mistakes that characterizes craftsmanship. However, this idea was challenged in the exhibition *The Power of Making* at London's Victoria and Albert Museum (Charny 2011). The works on display demonstrated a new digital designer-maker definition, whose process works between screen and reality with virtual and physical model making in an iterative process, and methods of production which effectively utilize mechanical and digital developments together. These works suggest an evolution of traditional crafts practice that is not divorced from its established principles.

The idea that craftsmanship and mechanized technologies are at odds is not the case: "Skilled artisans are developing unique businesses that focus on a particular technology, such as laser or water-jet cutting, digital printing and prototyping, resin prototyping, metal casting and pressing or computerized textile weaving" (Cochrane 2007, p. 14). This fusing of practice suggests that crafts can include digital technology in the making process, so the important elements are rather the connection, the meaning and the values embedded in the forms. Mass produced work is generic by definition. In responding to the human need for individuality and expression, the designer-maker needs to help the individual to understand the value in the product. This has been based on a dialogue, such as a text explaining a work of art that includes, for example, social commentary. This has been a challenge to communicate and does not obviate the costs involved. Until the recent development of web communication, this has been outside the normal retail channels of designer-makers, whereas now it is much simpler to communicate their values to a wider market. Craftspeople have comfortably integrated mechanization into their practice throughout the years and, contrary to the idea of digital technology distancing design from making, the reality is that it is bringing it back together (Loy and Canning 2013). This is happening on two levels. On one level, it is allowing for commercial production to proceed, driven by highly skilled craft skills, where the output can be repeated, customized or individualized, making skilled production more commercially viable. Second, it is reconnecting people to the construction of products. Improving the connection of people to objects supports the ideals of sustainable design, placing 3D printing at the forefront of a new 'Maker' movement that positively changes people's relationship with the products that populate their world.

6 Creating Balance

Diegel, referencing Sosa and Gero, suggests that "design practitioners, through their roles in shaping the future, are viewed as being able to promote change in society, especially around unsustainable behaviours" (2008) and highlights the arguments of Whiteley that designers have a "moral and ethical obligation to be responsible for their designs, and the social and environmental impacts of their work" (Diegel et al. 2010, p. 68). David Trubridge, as a designer-maker, demonstrates this approach, with sustainability principles—including sociocultural sustainability—clearly embedded in his work. Alongside these sustainability drivers, digital fabrication technologies are embraced, with CNC routering an integral part of the making process. His work provides a blueprint for maximizing the opportunities provided by a digital fabrication technology as it has resulted in a changed business model, rather than being just additional technology in his workplace. Trubridge has established a workshop-based production facility, based on a designer-maker ethos informed by conventions of craft, and embedded with cultural and environmental sustainability drivers. He articulates the drivers both through the design and fabrication of the work and through the workshops and lectures he delivers around the world. In answering the question "Why design?", posed at the Cooper Hewitt, Smithsonian Design Museum, Trubridge demonstrates this integrated approach of embedding cultural, social and environmental concerns in his designs: "To provide cultural nourishment, to tell stories, to reach people emotionally and spiritually; the objects are a vehicle for the nourishment we so badly lack in all the pragmatic and consumer stuff we are surrounded with. Most of the talk about design's role in helping us solve our current problems is still very rational, left brain thinking…but I strongly believe that design also needs depth, a different dimension that nourishes us culturally as well as providing us with tools. This is how it has been in the past throughout all cultures…and the other reason I design is to recreate that vital connection to nature that we have lost so much, living in insulated cities. We desperately need to rebuild that connection so as to value nature because it is what gives us life: fresh water, clean air and food. If we go on destroying it in the way we are now, it is us that will suffer the most, along with all the other species we make extinct" (Shelly 2010). This attitude aligns with the discourse on the role of the designer-maker in sustainability discussed in this chapter.

Trubridge states his intention is to design with the environment in mind, not only with nature as a source of inspiration to his designs, but extending into material choices, fabrication and distribution methods. He contemplates the meaning of the 'cultural designer' in practice and the complex issues and context designers work within today, including an overload of unnecessary design. His response to this is to create designs using local materials efficiently which communicate stories and embed rituals into their forms and use. Trubridge draws heavily on nature and cultural myths for inspiration alongside traditional hand-making techniques, such as steam-bending and basket-making, in his works. These influences are re-interpreted into abstract and repeated forms using systems-based construction techniques that utilize digital technologies in their manufacture. Story telling is a key element in his designs, particularly the Maori culture of New Zealand where he is based, connecting his work to place not only through his material choices, but through his engagement with cultural heritage. Trubridge references very specific geographical and cultural locations to represent unique experiences of place. This approach is supported by Lippard, who states that place is: "latitudinal and longitudinal within the map of a person's life. It is temporal and spatial, person and political … it is about connections, what surrounds it, what formed it, what happened there, what will happen there" (Lippard 1997, p. 7). An example of Trubridge's work that demonstrates this, called *Above Eye Level*, was shown at the Milan design fair (2015) as an installation commenting on rising sea levels. Trubridge describes it as "a story of how sea level rise is affecting some very small, marginal communities on pacific islands" (Trubridge 2015). Trubridge sought to highlight the impact that the behaviour of developed nations (and their unsustainable practices) is having on their pacific neighbors, with several Islander communities experiencing land loss caused by climate change. The form shown in Fig. 1 was inspired by a traditional boat of the Papua New Guinea people, a Thofothofo: "the intent of the form to evoke a skeleton that is washed up on the beach, a remnant of a culture that has been left behind" (Trubridge 2015).

For designer-makers such as Trubridge, extending practice into utilising additive manufacturing technologies could align with the use of CNC in providing opportunities to create wider ranges of customized outcomes. Additive manufacturing began as a prototyping tool, so for the designer-maker it was mostly useful for where it could replicate the constraints of a manufactured product, in particular injection moulding, prior to investing in tooling. This was more relevant to large-scale manufacturers because it was relatively expensive, and because investing in tooling was rarely worthwhile, as it required a relatively large run to recoup the initial costs. However, as the technology has shifted in the last 10 years from a prototyping technology to an end-use, direct manufacturing technology, it has become far more relevant to the designer-maker. The ability to produce more complex geometries in end-use products is as relevant for the individual designer as it is for larger manufacturers. However, the other benefits for the designer-maker have a different emphasis. For example, manufacturing, topological optimization for parts that reduce weight, as in the Airbus A380 metal brackets, is a major feature. For the designer-maker, maximising the raw materials would be a relevant feature in line with sustainability, but their objects are not produced en mass and it is not in itself a major issue. Rather, for the designer-maker working on 'connection to people and place' with site-specific products expressing values and telling

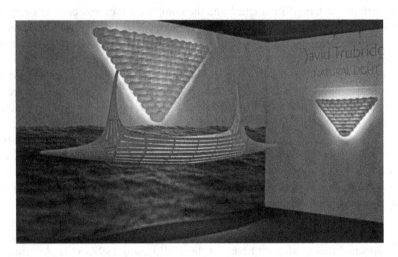

Fig. 1 Above Eye Level by David Trubridge

stories, the ability to create one-off or short run objects would most likely be seen as the key factor. An example of how it could be used would be in the design and 3D printing of individual, site-specific hardware, such as metal hinges and handles—a challenge to produce as one-off designs using alternative technologies. It would also be possible for designer-makers such as Trubridge to use 3D printing to create customized connectors in design for disassembly that would align with his current practices using CNC routering. However, current barriers to adoption could include the transport miles involved in having objects printed off-site and sent back to the workshop, such as would be necessary with 3D printing metal products (adding to the embodied energy in the product), and the inability to integrate local materials into the process. These factors would be mitigated by the fact that metal hardware would probably already need to be brought in, and by the refinement of wood-based filaments for fused deposition modelling utilising local wood off-cuts unsuitable for conventional making processes.

However, whilst the challenges for integrating 3D printing into production for designers such as Trubridge are still an issue, for other designer-makers it is the opposite, in that it is only through the technical possibilities provided by the technology and the disruptive business practices that it provides that they are now able to operate. 3D printing is allowing a new paradigm to develop that supports the work of individual designer-makers around the world, enabling them to produce high quality, highly invested designs with the marketing and production framework to be sold at prices that can support their individual existence, without compromising their practice. This is a game-changer for designer-makers around the world.

Essentially, the 'Master' craftsperson has been limited by his or her ability to create sufficient high end product without compromising his or her output to make a living. The reality is that investing a considerable length of time in the development of a single object that requires a high level of skill and embodies ideas that have been nurtured over that extended time results in an expensive outcome. This limits the market so narrowly that few traditional craftsmen can survive. However, the work of recent digital designers, such as Bathsheba Grossman, utilizes 3D printing to challenge that paradigm. Grossman developed the now iconic Quin lamp. Grossman has a background in sculpture, and her work supports the notion that 3D printing is not working against traditional making practices, but is an extension of them. The Quin lamp took months of work to develop and would not be a commercially viable product except that it can be repeatedly printed on demand.

In addition to supporting the commercial viability of individual designer-makers, 3D printing allows for 'master craftspeople of 3D printing' to emerge, as discussed in the work of Loy and Canning (2015). Grossman exemplifies this new tradition in that her designs demonstrate an understanding of the technology, resulting in objects impossible to create using conventional technologies, including hand-making. "These two sides, on the one hand virtual creativity and designing and on the other the industrial technologies, have become bound to each other, although not exclusively, by offering almost unbelievable possibilities that cannot be achieved in any other way. They enable designer-makers to generate forms digitally and make them tangible" (Shillito 2013, p. 16). Understanding how to maximize the technology requires an understanding not only of the modelling software but also of the constraints and opportunities of working with the technology to achieve an outcome unique to 3D printing.

The work of David Haggerty demonstrates this 3D printing 'master craftsmanship'. Haggerty works seamlessly between the virtual and physical in creating his work, pushing the boundaries of what is possible to produce. Even using a simple, low cost, dual filament printer, Haggerty's work demonstrates the creative potential of the technology as shown in Fig. 2.

Haggerty's metal work is part of a growing movement changing conventions in jewellery design. Haggerty uses 3D printing both for direct printing and for lost wax casting, mixing traditional practices with digital innovation. In Fig. 3 the first work was directly 3D printed in stainless steel electroplated with nickel, and the second was 3D printed in wax, then cast in bronze.

In addition to supporting the commercial work of an individual designer-maker such as Haggerty to add to the evolving community of craft practice, the flexibility of on-demand production also allows designers to design responsively. This can facilitate the production of localized solutions and the creation of community responsive solutions to local issues. The facilitation of designer-makers through 3D printing provides an opportunity to reconsider society's views on, and approaches to, consumption by providing opportunities to design for specific needs, and, with the ease of individual production, potentially to develop products with materials suitable for use (bio, ephemeral, permanent). Through the opportunity for customization, greater longevity of products (spreading embodied energy) can potentially be created if customers see value in truly individualized objects

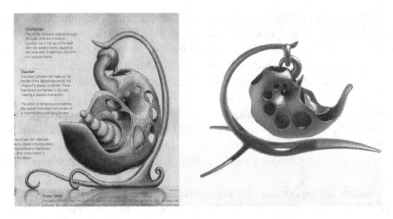

Fig. 2 David Haggerty's drawing and 3D print of 'Fractilus' design

Fig. 3 Examples of jewellery designed by David Haggerty

(meeting the requirements and needs of the user, rather than forcing the user to adapt to a generic object—potentially creating obsolescence in its generic solution). If so, it is up to society to start to change the consumer paradigm around this opportunity. For 3D printing, this is needed to ensure that the benefits of the technology to support more thoughtful products that uphold the current aspirations and values of society around sociocultural sustainability rather than homogenization are not swamped by the mass of low-end products being created. The reconnection of people to their needs to be supported and the framework for understanding qualities of design that utilize 3D printing are still being explored.

7 Digital Adaptation

The drivers underpinning an intellectual approach to craftsmanship endure over time, adapting to changing media and changing societal concerns. The discourse remains fresh because of the changing environment and the changing political landscape worldwide and how issues inform values and aspirations for a positive, sustainable global community. 3D printing is a significant challenge to conventional ideas of making that support the ideals encouraging consumerism based on connection to people, places, environment, cultural identity and diversity, yet it has the potential to reconnect where mass production has disconnected and that is in line with current thinking driving sustainability. Additive manufacturing technologies for direct manufacturing are still in their infancy, but developing associated tools, such as scanning and modelling software using haptics suggests that this tactile connection grows with 3D printing rather than be broken. The likelihood is that the more artisans are involved in using the technology, the more new ways of making that draw on digital technology could be developed. It is interesting to note that the mechanized Jacquard loom in 1801 was invented by the son of a master weaver and drew on the intricate knowledge of the process that Joseph-Marie Jacquard had.

3D printing for direct manufacturing began in engineering and medical applications, but has moved into creative applications as end-use materials and technologies have evolved. Because of this, it has developed across disciplines, creating links not normally there (Loy and Canning 2015). One of the advantages of this is that 3D printing is challenging the conventions of traditional making disciplines, allowing for new forms of practice. Examples in the MGX collection, such as the work of Grossman, illustrate that additive manufacturing technologies allow for new design thinking and also create cross-disciplinary collaborations, such as across fashion and engineering. Arguably, the most significant development in 3D printing and making has been in its enabling role in FabLabs (Gershenfeld 2005), and the development of the 'Maker' society. Whereas working in the computer environment involves the maker, the manufacturing of a computer-designed object at a distance is passive. However, there is a shift in behaviours and attitudes that comes from the immediacy of the desktop printers. On the face of it, the explosion of 3D printing of desktop objects by amateurs adds to the problems of the environment. The ABS and PLA objects printed by individuals not trained in design are unlikely to have an integrity of construction to create functional, responsible objects that add value to the material culture of communities. It is hard to argue that there is a quality outcome in the majority of objects being printed and displayed on 3D printing websites. Yet, if sociocultural sustainability depends on strengthening connections between people and objects, then the most fundamental factor with regards to the relationship between additive manufacturing technologies and sustainability is, ironically, the reconnection of people to making in a digital world. By reconnecting people to making using 3D printing, and involving them in the challenges in creating objects digitally, the process is educating them on construction, materials and aesthetics through the same experiential learning for tacit knowledge advocated by crafts commentators over the years.

In addition, by educating users on construction and making, the potential emerges for design for repair for products not seen since the move towards obsolescence in products following the economic motivators that drove product development practices after the Second World War. This fits with the current sustainability strategies of design for disassembly and product service system thinking proposed by theorists such as Hawken et al. (2010) and Ryan (2004), which are integral to the response of manufacturers following the introduction of EU legislation on the return of certain products at the end of life as the responsibility of the European manufacturer, and support changes in practices to meet the legislation on embodied energy and environmental impact being introduced in 2019.

3D printing should not be viewed in isolation but as part of much broader social developments inspired by the digital environment. In addition to connection to making and therefore objects that populate the environment, digital technologies, such as Web 2.0, have created a highly complex web of interactivity built on open source sharing and collaborations. By rethinking definitions of connection to community and what constitutes a community group, technically the digital world is providing new ways of addressing sociocultural sustainability to which 3D printing is integral.

Embracing the digital within a design and craftsmanship framework is resulting in new practices that align to sustainability strategies by supporting the creation of print on demand products that the end-users are involved in creating to an extent, and in which they are invested. This approach is converging the digital and the physical in an iterative process that connects the human and the technological. This has been supported by the rise of 3D haptic tools and touch-sensitive software tools to model 3D objects that are again shifting the skills involved with working in the making environment, but these advances are bringing the technical skills and traditional hand skills back together. There is a growing sophistication in the thinking around the relationship between the designer-maker and the digital craft that has developed since the turn of the century that works within the new environment and is aligned to supporting sociocultural sustainability. The work of Lionel Dean, founder of *FutureFactories*, creates a working practice that sustains the personal commentary of his work, but exploits the digital potential to provide what he terms an individualization of the design, rather than a customization. By making this distinction, Dean retains the authorship of his work, whilst collaborating in a co-design aspect with the user. Dean's work is part of a growing movement, as illustrated by the range of work in the publication Digital by Design, where the 'technology infused' designs describe how designer-makers, designers and artists are engaging creatively with the new digital tools with a maturing level of expertise. This work is at the forefront of merged practice with its value only beginning to be recognized more broadly: "People are not yet trained to value the labor that goes into coding a piece of software, for example or crafting an object with digital technology. Over time, our ability to recognize and value these aspects of digital design will hopefully grow, and most people will be able to distinguish complexity for the sake of complexity from genuine digital craftsmanship" (Warnier et al. 2014, p. 41). The user understanding of digital craftsmanship and its manifestation in products, such as those shown in the MGX collection, need to grow to promote the investment of users in the work to appreciate the expertise and investment of time and expertise involved in the same way traditional craftsmanship during the last century depended on the opportunity for the communication of the story of the product and the designer-maker. This does not change. If anything, digital communication improves the opportunity for that communication to take place, not only at the end of the process but during the design development phase.

8 Conclusion

Designers, by the current definitions discussed in this chapter, cannot avoid being involved in questions about what sort of society their work supports. For a genuinely complex sustainability approach to take hold, then the dominance of mass production needs to be reduced to balance worldwide patterns of living that are thoughtful and responsible. However, it is a challenge to maintain the ideals of craftsmanship and support the work of the individual designer-maker for sociocultural sustainability reasons, and yet work with, and respond to, current realities. A 'step-change' in thinking is frequently required to alter more radically practices after a particular idea has dominated.

Arguably, the most significant impact on designer-makers by digital fabrication and Web 2.0 has been the placing of the full production cycle into the hands of the designer-maker. This in many ways returns the practice back to preindustrial revolution times before the systemization of production and the subsequent separation of manufacturing and retail disconnected making and consuming. Web 2.0 and digital fabrication are now providing the possibility for a reconnection between makers and users that was not possible in the last century, as it allows for the designer to exert control over the full production life cycle and, more fundamentally, change the relationship between the user and the products that populate their environment.

The introduction of CNC practice disconnected the hand and the material. However, the skill involved in working with the computer and in making by hand is secondary to the understanding of form as the expression of, and connection to, ideals, values, places, people and materials. Whilst some theorists may have a problem with the place of 3D printing in a craft discourse, practitioners have addressed these issues previously with earlier digital technologies. In the work presented as part of the exhibition on *Pioneering the craft: CNC technology in the creation of furniture* displayed during the Furniture Society 2004 conference in Savannah, the correlation between the introduction and concerns over the use of computer aided design and production and the initial trend towards using table saws, band saws, etc. early in the twentieth century were highlighted. There was concern at the time about the values that might be lost, whilst recognising that the technology allowed for the accurate cutting and shaping of parts, more cost effective batch production for identical parts and the creation of "shapes, surfaces and textures that traditional tools and techniques could either not practically make, or not make at all" (Furniture Society 2004).

There is a danger in romanticising craft practice, partly in 'idealising' it out of viable economic competitiveness, but, more than that, there is danger of making it less relevant by imposing rules on practice based on outdated ideas. The issues involved in a discussion about designer-makers and 3D printing are part of a much wider shift towards digital practice, and the reality is that if the fundamental ideas of connection remain in place, then the craftsperson should be able to use all means possible to enhance his or her practice. Equally, their work should more likely resonate with a population if it draws on the media used collectively at that time. To support the reconnection to material, place, environment and culture, advocated as a sustainability strategy at this time, there needs to be support for designer-makers to respond. If mass production overwhelms individual designer-makers, then a significant part of the sociocultural sustainability imperative is lost. Tannen argues that: "craft knowledge survives through its practitioners. In a culture dominated by distributed systems of manufacture, uniformity, and consumerism, we must take care that this personal know-how continues to exist, and is manifested in the wonder of making. Though the crafts are marginalized today in almost every conceivable way, I believe that they have never been more important to our culture, both in process and product, as an expression of important values that are at risk" (Tannen 2005, p. 14).

It is not technology alone that drives unsustainable solutions. Designers, makers and ultimately consumers can consider when using these technologies how longevity and cultural values are embedded into the objects being created. Changes in sociocultural norms, particularly in relation to sustainability—cultural, environmental, economic and social—impact on sustainable practice. They are interrelated. Key is how the values of the maker practice and the perceived value of the crafted artefact can be transferred to a digitally fabricated object.

Designers are operating in an increasingly globalized society, with production models and processes distributed globally. It is a unique point in time, where the main means and methods of creation are producing aesthetics and forms that are increasingly homogenized. In contrast, 3D printing allows for an unprecedented level of customization and the creation of complex forms. Invested design shifts the role of the user to what could be better termed an owner. The idea of ownership determines a sense of responsibility, both for the production of products and for their maintenance. This could help to foster behaviours that encourage and protect diversity, rather than diminish it. Projecting forward, the development of digital and real-world outcomes in terms of communication, design and making continues to merge as technology advances. The stories behind the objects are more easily told to a wider audience. However, more fundamentally, the broader understanding of working with the merging digital and real-world environments is growing. This allows for the development of strategies for designer-makers to assimilate digital developments, such as 3D printing. Ideas informed by a technical understanding of both the digital and the physical making environment provide opportunities to develop approaches to design that maintain sociocultural sustainability.

The reality is that the digital methods of creating 3D forms are still in their infancy, and it is only recently that designer-makers are starting to combine effectively their values-based practice with the new technology. The challenges in terms of creating connections and responding to sustainability drivers confronting the designer-maker, combined with the tools of the digital revolution, are new in relation to the integration of the technology with maker practice, but the ongoing dialogue is the same. Practice is inevitably changed by the means of production now available, just as designs and practice changed with the availability of CNC routering to the designer-maker. However, more than ever, the sustainability imperative is forcing a rethink of making, production systems and consumption of natural resources. The reflection and expression of much broader social ideas and providing a means to influence values is more important than before. The rethinking of production systems to reduce waste and improve embodied energy targets for working within the constraints of what the planet can produce and maintain points towards the small-scale designer and invested products, yet they have to be able to compete with low-cost mass production. However, digital technologies, and in particular 3D printing, may now be providing answers to supporting sustainability ideals not previously realistically possible to implement.

References

Carson R (2002) Silent spring anniversary edition. Houghton Mifflin Company, Boston

Charny D (ed) (2011) The power of making: the importance of being skilled. V&A Publishing and the Crafts Council, London

Cochrane G (2007) Smart works: design and the handmade. Powerhouse, Sydney

Diegel O, Singamneni S, Reay S, Withell A (2010) Tools for sustainable product design: additive manufacturing. J Sustain Dev 3(3):68–75

Elkington J (1999) Cannibals with forks: The triple bottom line of 21st century business. Capstone, Oxford

Fuad-Luke A (2009) Design activism: beautiful strangeness for a sustainable world. Earthscan, London

Fletcher K, Grose L (2012) Fashion and sustainability. Laurence King, London

Gershenfeld N (2005) Fab: the coming revolution on your desktop—from personal computers to personal fabrication. Basic Books, New York

Gibson I, Rosen D, Stucker B (2015) Additive manufacturing: 3D printing, rapid prototyping and direct digital manufacturing, 2nd edn. Springer, New York

Gunn W, Otto T, Smith R (eds) (2013) Design anthropology: theory and practice. Bloomsbury Academic, London

Hawken P, Lovins A, Lovins H (2010) Natural capitalism: the next industrial revolution, 2nd edn. Earthscan, London

Hickman L (2009). Slow food: have we lost our appetite? The guardian. http://www.theguardian.com/environment/2009/feb/04/slow-food-carlo-petrini. Accessed 27 Aug 2015

JamFactory (2013) Designing craft / crafting design: 40 years of JamFactory. JamFactory, Norwood Aus

Kawamura Y (2012) Fashioning Japanese subcultures. Berg, London

Lippard L (1997) The lure of the local: senses of place in a multicentered society. New Press, New York

Loy J, Canning S (2015) Clash of cultures. In: Connor A (ed) Manual of research in creative technologies. IGI Global, Hershey Pennsylvania

Loy J (2014) eLearning and eMaking: 3D printing blurring the lines between the physical and the digital. Educ Sci 4:108–121

Loy J, Canning S (2013) Reconnecting through digital making. In: Pandolfo B, Park M (eds) Make. Industrial design educators network, no 2, pp 12–21

McDonough W, Braungart M (2002) Cradle to cradle. North Point Press, New York

Makepeace J (1995) Makepeace: spirit of adventure in craft and design. Conran Octopus, London

Norman D (2013) The design of everyday things: revized and expanded edition. Basic Books, New York

Ryan C (2004) Digital eco-sense: sustainability and ICT—a new terrain for innovation. Lab 3000, Melbourne

Shelly K (2010) Designer David Trubridge answers why design? Cooper Hewitt. http://www.cooperhewitt.org/2010/06/08/designer-david-trubridge-answers-why-design/. Accessed on 19 Aug 2015

Shillito A (2013) Digital crafts: industrial technologies for applied artists and designer makers. A&C Black, London

Swartz M, Yair K (2010) Making value: craft and the economic and social contribution of makers. Crafts Council, London

Tannen R (2005) Furniture makers exploring digital technologies. The Furniture Society, Asheville NC

Trubridge D (2015) So far. Potton & Burton, Port Nelson NZ

Walker S (2006) Sustainable by design: explorations in theory and practice. Routledge, London

Warnier C, Verbruggen D, Ehmann S, Klanten R (eds) (2014) Printing things: visions and essentials for 3D printing. Gestalten, Berlin

Additive Manufacturing and its Effect on Sustainable Design

O. Diegel, P. Kristav, D. Motte and B. Kianian

Abstract 'Sustainability' is an emerging issue that product development engineers must engage with to remain relevant, competitive and, most importantly, responsible. Yet, on examining the term 'sustainable', a plethora of definitions emerges, many of which are contradictory and confusing. This confusion and a general lack of understanding means that sustainability often gets relegated to an afterthought or a buzz-word used on marketing material, no matter how 'sustainable' the product actually is. The role of the 'sustainable' product developer is to look for new opportunities to design products that minimize harmful effects on the environment and to seek to develop environmental, social, and economically beneficial product solutions. The advent of additive manufacturing technologies presents a number of opportunities that have the potential to benefit designers greatly and contribute to the sustainability of products. Products can be extensively customized for the user, thus potentially increasing their desirability, pleasure and attachment—and therefore longevity. Additive manufacturing technologies have also removed many of the manufacturing restrictions that may previously have compromised a designer's ability to make the product they imagined which, once again, can increase product desirability, pleasure and attachment. As additive manufacturing technologies evolve, design methodologies for lightweighting, such as topology optimization, become more advanced, more new materials become available, and multiple material technologies are developed, the field of product design has the potential for great change. This chapter examines aspects of additive manufacturing from a sustainable design perspective and looks at the potential to create entirely new business models that could bring about the sustainable design of consumer products. It first gives a brief literature review both on sustainable product development and on additive manufacturing, and then examines several case study products that were made with additive manufacturing. It concludes that there is a likelihood that

O. Diegel (✉) · P. Kristav · D. Motte · B. Kianian
Product Development Research Group, Faculty of Engineering,
Lund University, Lund, Sweden
e-mail: olaf.diegel@design.lth.se

© Springer Science+Business Media Singapore 2016
S.S. Muthu and M.M. Savalani (eds.), *Handbook of Sustainability
in Additive Manufacturing*, Environmental Footprints and Eco-design
of Products and Processes, DOI 10.1007/978-981-10-0549-7_5

additive manufacturing allows more sustainable products to be developed, but also that more quantifiable research is needed in the area to allow designers to exploit better the features of additive manufacturing that can maximize sustainability.

Keywords 3D printing · Additive manufacturing · Design freedom · Planned obsolescence · Product attachment · Sustainability · Sustainable product design

1 Introduction

The past decade has seen a surge in awareness for environmental conservation and the preservation of the Earth's natural resources and environment. Sustainability is rapidly emerging as an issue that designers and engineers must engage with and embrace to survive in a more sustainability conscious world. Indeed, sustainability is now taught as an integral part of many design and engineering degrees, recognizing the growing acceptance of the role sustainability has to play in the development of our futures. Yet, on examining what is meant by 'sustainable' products, a plethora of definitions and methodologies emerge, many of which contain fundamental omissions or contradict each other. This confusion means that sustainability often gets relegated to being just a buzz-word used on marketing material no matter how sustainable the product actually is. Part of the challenge for product development engineers and designers is to move beyond the "hype" and to engage in design activities with the level of integrity that our futures require.

Product development and design practitioners, through their roles in shaping the future, are viewed as being able to promote change in society, especially around unsustainable behaviors (Sosa and Gero 2008). In "Design for society", Whiteley (1993) argues that designers have a moral and ethical obligation to be responsible for their designs and the social and environmental impacts of their work. Whiteley (1993) follows the writings of others (i.e., Papanek 1985) to reveal a lack of values and ambition in the juxtaposition between design and consumerism. Consumer-led design is so prevalent that it appears as a "natural and inevitable aspect of our society" (Whiteley 1993). For design to change, the role and values of design, as well as the relationship of design with society, need to change. This may come from a reflection as to whether design is merely a servant of industry or can inform through intelligent thought and action, while contributing to the global ecological balance (Whiteley 1993).

The design community is consequently in a state of transformation. Designers have responded to the growing issues around social and environmental issues by developing concepts and frameworks such as eco-design, sustainable design, and numerous related iterations (Sherwin 2004). These concepts are centered on ideals of acknowledging ecological limits and demonstrating responsibility, and increased contribution to society and the environment (Sherwin 2004). Within the context of product design, approaches to sustainability generally fall between two broad areas: eco-design and sustainable design (Tischner and Charter 2001; Sherwin 2004). Although these methods are essential, and incredibly helpful, in guiding designers through the process of designing sustainable products, most of them do not explore or capture the potential for new and developing technologies to help support the product development process. Many sustainable design attempts appear to be "one-off" or experimental designs. Although this process is an essential part of the development towards understanding the role of design in developing true sustainability, it also demonstrates the uncertainty surrounding how the principles of sustainability can be successfully incorporated into mass-produced everyday consumer items.

In this chapter, Sect. 1.1 deals with various aspects of product sustainability including planned obsolescence, and a number of sustainable design approaches that have been used over the years. Section 2 then deals with the relationship between design quality and product longevity, and proposes the hypothesis that better designed products result in customers keeping them for longer, thus increasing product sustainability. Section 3 then gives some background information on additive manufacturing and some of the advantages it offers over conventional manufacturing. It then proposes additive manufacturing as one of the tools that allows product designers to develop better products that increase customer attachment and thus sustainability. The chapter then concludes with some thoughts on what further research is needed in this area in order to quantify better the effects of additive manufacturing on good design and increased product longevity.

1.1 Planned Obsolescence

Of the world's nearly 7 billion population, about 1.7 billion people now belong to the "consumer class" (Halweil et al. 2004) with lifestyles devoted to the accumulation of non-essential goods, characterized by a desire for bigger houses, more cars, and more consumer goods. Nearly half of these global consumers reside in developing countries. There are 240 million consumers in China and 120 million in India and these are the markets with the greatest potential for expansion.

In the context of environmental sustainability, the consumerist's "throw away" mentality has had a strongly negative effect on the planet in terms of pollution levels, water supplies, natural habitats and ecosystems. This throw away mentality extends from disposable cameras to other cheaply made goods with built-in product obsolescence that have little consumer attachment.

Planned obsolescence, or built-in obsolescence, is the process of a product becoming obsolete or non-functional after a certain period of use in a way that is planned or designed by the manufacturer. Planned obsolescence has potential economic benefits for a producer because the product fails and the consumer is under pressure to purchase again.

For an industry, planned obsolescence stimulates demand by encouraging customers to purchase again if they want a functioning product (Bulow 1986; Waldman 1993). Built-in obsolescence exists in many different products, from vehicles to light bulbs to proprietary software.

Design for disposal encourages throwing away sophisticated and energy-expensive items. Some brands of inkjet printers, for example, incorporate the print head technology within the cartridge so that it must be discarded when the ink container is empty. Separation of a simple ink container from a permanent print head would mean that the energy expensive part of the printer could be used again. Many products fall into this category—use for a relatively short time and then discard.

The concept of planned obsolescence is not new (Packard 1978). Planned obsolescence was first developed in the 1920s and 1930s when mass production had opened every minute aspect of the production process to exacting analysis, and the prime goal of most businesses was only centered on economic sustainability. In the 1930s an engineer working for General Electric proposed that increased sales of flashlight lamps could be achieved by increasing their efficiency while shortening their life. Instead of lasting through three batteries, each lamp should last only as long as one battery, thus forcing the consumer to purchase more lamps. In 1934, speakers at the Society of Automotive Engineers also proposed limiting the life of automobiles in order to increase sales (Packard 1978).

By the 1950s, planned obsolescence had become routine and designers worried over the ethics of deliberately designing products of inferior quality. The conflict between profits and design objectives was apparent. The fear of market saturation seemed to require such methods to ensure a prosperous economy, yet the consumer was being sold inferior products which could have been made more durable for little extra cost.

Today there are, quite sadly, too few products that are not designed with planned obsolescence in mind. It is evident that planned obsolescence and environmental sustainability are in almost direct conflict. Though drastic changes in consumer culture are beyond the scope of this chapter, designers can certainly play a role in designing products that go beyond planned obsolescence. The advent of additive manufacturing (AM) offers some potential in this area as it allows designers to, potentially, produce products without the compromises they are forced into by conventional manufacturing and this, again potentially, may allow the creation of products that are more desirable and therefore, as is argued later in this chapter, more sustainable.

1.2 Sustainable Design Approaches

It is of interest to examine the literature on sustainable design approaches in order to help us to understand better what factors affect product sustainability and, hopefully, to engender some new ideas into how emerging technologies can be used to stimulate more sustainable product design. This section of the chapter describes a few of the current approaches towards sustainable design, and explores the relationship between design quality and product longevity and sustainability. It also proposes the potential for new manufacturing technologies, such as AM, to play a greater role in the design of sustainable products.

1.2.1 Eco-design

Eco-design aims to incorporate environmental issues into every product with the aim of minimizing its impact on the environment (Tischner and Charter 2001). It is often marketed through the economic gains resulting from the cost reduction associated with the corresponding gains in efficiency. Eco-design considers the environment at each design and manufacturing stage so that each stage makes the smallest environmental impact throughout the product's life (Glavic and Lukman 2007). In most real-world applications, eco-design generally manifests itself in the production stages through attempts to use materials with low environmental impact. Environmental impacts after the product have been sold and before recycling are, however, generally not taken into account (Ljungberg 2007).

Eco-design tools can, broadly, be divided into tools for environmental assessment and tools for environmental design (Le Pochat et al. 2007). In terms of environment assessment tools, life cycle analysis (LCA) is probably one of the most commonly used tools to assess the environmental impact of a product (Ciambrone 1997; Wenzel et al. 1997; Sherwin 2004). Because of the nature of LCA, however, the environmental impact of a product is usually evaluated at the later stages of the design process after most of the design decisions have been made, rather than during the planning and conceptual stages, which are the ones that have the greatest effect on true environmental impact (Sherwin 2004; Kobayashi 2006). This reduces the effectiveness of LCA as improvements that may benefit the environment cannot be introduced until subsequent product iterations (Sherwin 2004). The LCA approach used by many companies also generally focuses on the technicalities of the design, such as the optimization of material components. Designers are consequently often not involved in this part of the process, which is left to engineers, thus causing LCA to cover only a small part of the product development process (Abukhader 2008). Other assessment tools, allowing for simplified LCA analysis, such as the Eco-indicator 99 (Goedkoop et al. 2000), allow an earlier assessment, but at the cost of a precise assessment (Dreyer et al. 2003).

Tools for environmental design consist mainly of manuals, checklists, or guidelines (e.g., Telenko et al. 2008), or descriptions of existing green products as a source of inspiration, such as the *Eco-Design Handbook* (Fuad-Luke 2009). They allow the assessment of the environmental earlier in the design and development process.

Many methods and tools have been developed for eco-design, but many of them are the results of smaller research projects or consulting projects in collaboration with specific companies or industries. There is a need for consolidation of these tools, better integration into the design and development process of companies (Le Pochat et al. 2007), and, above all, further testing (Vallet et al. 2013).

1.2.2 Sustainable Design

Though sustainable product design is an extension of eco-design, it goes well beyond the principles of eco-design. While encompassing elements of eco-design,

Fig. 1 Triple bottom-line
product design (Diegel 2015)

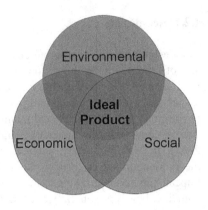

sustainable design also incorporates economic imperatives, ethics, and other socio-economic dimensions of sustainability, and uses ecological principles as methods of designing, thus aiming for "triple bottom-line" solutions (Tischner and Charter 2001; Sherwin 2004).

Triple bottom-line, as shown in Fig. 1, divides sustainability into three areas: environmental, economic, and social sustainability. An ideal product is one which maximizes all three areas in that it is good for the environment, is profitable for the company, and improves society.

Economic sustainability is relatively easy to measure as it is, to a large degree, easy to quantify. Social sustainability is somewhat more difficult to measure because of the intangible nature and subjectivity of many of the factors that are deemed of benefit to society. Does a gun, no matter how well designed, have a positive or negative impact on society? Environmental sustainability, from a product point of view, is also difficult to quantify as, to get a true understanding of a products' impact, one has to look at the entire life cycle of the product which can become quite a complex endeavor. Even using statements such as "good for the environment" as used in the ideal product description above, can be misleading. Is any product, in fact, good for the environment, or is the role of the product developer simply to try and minimize the negative impacts of a product on the environment?

As sustainable design is a discipline that has only begun to grow in importance over the last 20 years, there are currently few models for implementing it in practical product design projects (Tischner and Charter 2001). As an attempt to resolve the natural tensions between the three areas described above, the literature on sustainable design presents five common models to help in the management of sustainable product design. It should be noted, however, that four models of the five tend to focus on environmental and economic issues and do not attempt to address the wider social and ethical issues of the product (Tischner and Charter 2001), and none of them address design quality as a factor that affects the longevity of a product, and therefore its sustainability.

1.2.3 Road-Map for Sustainable Product Development

Waage (2007) presents a four stage 'road-map' to guide product designers and companies through the complex issues of sustainability:

- The first stage is to understand the sustainability context for the product
- The second stage is the exploration of the sustainability issues pertinent to the product in question
- The third stage sees this exploration being defined, refined, and assessed to identify the best and most sustainable solutions
- The final stage sees the process of product implementation followed by customer feedback, and product impacts are monitored over time.

1.2.4 Cyclic-Solar-Safe Principles

Datschefski (2004) proposes five sustainable design elements developed to mimic plant and animal ecosystems in order to maximize the use of the finite resources at our disposal, while maximizing human happiness and potential. The five basic principles are described as follows:

- *Cyclic*: the product should be made from organic materials and be recyclable and compostable. If it is mineral based, it should be cycled continuously in a closed loop system.
- *Solar*: solar or renewable energy should be used during manufacture and product use.
- *Safe*: the manufacture, use, and disposal should be non-toxic and not interfere with, or disrupt, ecosystems.
- *Efficient*: a product's manufacture and use should use 90 % less material, energy, and water compared with a 1990s equivalent.
- *Social*: the manufacture and use of a product should not impact on basic human rights or natural justice.

1.2.5 Framework for Strategic Sustainable Development

The Framework for Strategic Sustainable Development (FSSD) is an overarching framework designed to take a full systems perspective with respect to achieving a sustainable outcome. It is commonly referred to as The Natural Step as this is the consulting group which utilizes the FSSD. The FSSD views the triple bottom line as a nested system approach where the economy is nested within the social system and which is in turn nested with the ecological system. It defines success as the meeting of eight design constraints or Sustainability Principles. These Sustainability Principles have been created to from the mindset of the most upstream methods of destroying our ecological and social systems, Thus when used as design constraints during product development they ensure that you design a product which does not undermine the ecological or social system's ability to self-renew. It should be noted that these Sustainability Principles are designed to be used in long-term strategic planning processes and so products design using this approach may not be in full alignment with the sustainability principles from the start but may work well as a process of refinement as new technology and material become available, thus ensuring a stepwise and financially feasible approach to long term sustainability (Missimer 2015).

The Sustainability Principles have recently been updated to include a more robust definition of social sustainability and are as follows (Missimer 2015). In a sustainable society, nature is not subject to systematically increasing:

- Concentrations of substances extracted from the Earth's crust
- Concentrations of substances produced by society
- Degradation by physical means

In a socially sustainable society, people are not subject to structural obstacles to the following:

Strategic Sustainability Principle 1: *health*

This means that people are not exposed to social conditions that systematically undermine their possibilities to avoid injury and illness, physical, mental, or emotional, e.g., dangerous working conditions or insufficient wages

Strategic Sustainability Principle 2: *influence*

This means that people are not systematically hindered from participating in shaping the social systems they are part of, e.g., by suppression of free speech or neglect of opinions

Strategic Sustainability Principle 3: *competence*

This means that people are not systematically hindered from learning and developing competence individually and together, e.g., by obstacles for education or insufficient possibilities for personal development

Strategic Sustainability Principle 4: *impartiality*

This means that people are not systematically exposed to biased treatment, e.g., by discrimination or unfair selection to job positions

Strategic Sustainability Principle 5: *meaning-making*

This means that people are not systematically hindered from creating individual meaning and co-creating common meaning, e.g., by suppression of cultural expression or obstacles to co-creation of purposeful conditions

These are underpinned by the following scientific principles:

- Total energy in a system remains constant (first law of thermodynamics)
- Matter and energy tend to disperse (lay of entropy)

- Society consumes quality, structure, or purity of matter, not molecules
- Increases in material quality on Earth are almost entirely the result of photosynthesis
- Society is a self-organizing complex adaptive system

1.2.6 Cradle to Cradle

Cradle to cradle is a relatively recent design framework. Similar to biomimicry and the natural step, cradle to cradle is a framework inspired by looking at natural systems (McDonough and Braungart 2002; McDonough et al. 2003; Braungart et al. 2007). A core tenet of cradle to cradle is a critique of the use of 'eco-efficiency' as a driver for developing environmentally benign products and systems. McDonough and Braungart argue for strategies such as doing more with less and focusing on maintaining or increasing economic outputs while decreasing the impact on ecological systems. Zero emission, which aims for maximum economic output with zero environmental impacts, is the ultimate goal of such approaches and represents a breakdown of the economic and ecological relationship (McDonough and Braungart 2002; Braungart et al. 2007). Braungart et al. (2007) describe eco-efficiency as:

- A reactionary approach that does not address the need for fundamental redesign of industrial material flows
- Being inherently at odds with long-term economic growth and innovation
- Not effectively addressing the issues of society

They regard eco-efficiency as being based on the assumption of cradle to grave material flow that transforms resources into waste, which is then buried in the Earth as a graveyard. Ultimately they regard eco-efficiency as "less bad is no good" (Braungart et al. 2007).

If one examines environmental sustainability independently from the other two triple bottom line sustainability elements, it could be argued that any physical product is, in fact, bad for the environment as it requires resources to get manufactured, distributed, and then disposed of at the end of its life cycle. There are few, if any, existing hardware products that, if one analyzes them from cradle to grave, or cradle to cradle, have a positive impact on the environment (Zafarmand et al. 2003; Sherwin 2004). It could therefore be said that part of the role of the 'sustainable' product designer is to design products that, while maximizing their economic and social impact, minimize their harmful effects on the environment.

1.2.7 Circular Economy

The circular economy concept presents five innovative business models and ten disruptive technologies which enable them, at both macro and micro level, to create value chains and foster decoupling growth from degradation of natural resources. It should be noted that the intention of creating circular value chains is not new. As described earlier in this chapter, cradle to cradle, or other concepts such as industrial ecology and biomimicry, explored and promoted similar concepts decades ago. However, what makes circular economy different and more implemented are the disruptive technologies. These disruptive technologies enable huge changes, which would have been impossible just a few years ago (Accenture 2014).

These business models and disruptive technologies are operating on longevity, renewability, reuse, repair, upgrade, refurbishment, sharing capacity, and dematerialization (Accenture 2014). The five circular business models are:

- Circular Supplies (integrating renewable or fully recyclable input, thus removing single use inputs)
- Resource Recovery (utilizing material and resources from previous waste streams)
- Product Life Extension (improving product life span through repair, upgrading, and reselling)
- Sharing Platform (collective shared use/access/ownership in order to improve utility rates of physical products)
- Product as a Service (reworking of original product sales to service models where physical ownership remains with the producer)

The ten disruptive technologies, which are essential to launch and operate the above business models, are classified into three categories of digital (information technology), engineering (physical technology), and hybrids of both digital and engineering. These ten disruptive technologies are (Accenture 2014):

- Mobile
- Machine to Machine technologies (e.g., Internet of things)
- Cloud
- Social
- Big Data Analysis
- Trace and Return Systems
- Additive Manufacturing (e.g., 3D printing)
- Modular Design Technology
- Advance Recycling Technology
- Life and Material Sciences.

2 Relationship Between Design Quality and Sustainability

It should be noted that the bulk of the literature on sustainable product design tends to focus on the technicalities of lowering the environmental impacts of material, resource and energy use. According to van Nes and Cramer (2003) and Vincent (2006), very little of this literature deals with 'design quality' as a factor in improving product longevity. By longevity we mean extending the useful life of a product, and therefore reducing the impact it has on the environment. Though there is a large quantity of literature on various aspects of design quality, and even on its importance to sustainability, there is little that states how it fits into the methodologies towards attaining sustainable product design (van Nes and Cramer 2005; Park 2005).

'Design quality' is a difficult term to define, as it is an area that is often regarded as subjective (Kemp and Martens 2007). If one examines quality from a sustainable product point of view, it could be argued that design quality has a direct effect on the longevity of a product (van Nes and Cramer 2005; Vincent 2006). Here we use design quality not just to mean the 'technical quality' of a product but also the less tangible 'desirability' of a product, 'pleasure of use' of a product, and the 'attachment' of a user to a product. Designers can stimulate desirability, increase pleasure, and deepen attachment by designing products that not only function better and are more aesthetically pleasing than comparable products, but are also tailored to suit better the individual needs of the user. Govers and Mugge (2004) argue that if an object is highly desirable its longevity is extended, and its negative impact on the environment is therefore reduced.

One could extend this argument to say that products which are so well designed that they become lasting 'objects of desire, pleasure, and attachment' are more sustainable because they do not get disposed of in the same way as lower quality designed products. From this, one could say that the E-type Jaguar, for example, is potentially more environmentally sustainable than a modern hybrid car because, if one looks at its complete life cycle, it performs superbly. This is because the quality of its design (and here we mean aesthetic design rather than engineering design) makes it such a great object of desire that it never gets scrapped as a conventional car might. It increases in value as time passes and is cherished by its owner, with great care being taken in its maintenance and, in all likelihood, it could last for several generations. This argument, however, is by no means universal, as some would argue that the poor engineering design of an E-type Jaguar make it a relatively poor performer in comparison with a modern car. The argument is merely an example of the argument one could make towards good design increasing the longevity of a product and, therefore, its sustainability.

Products that are experienced as "authentic" have a high value for their customers, and are hence kept longer. Gilmore and Pine (2007) have mapped the concept of authenticity into genres of how customers perceive authenticity and product value, each one corresponding to a particular form of offering. The most rudimentary form is, perhaps, the natural authenticity, which concerns commodities. Customers perceive things that exist in their natural state as authentic. The pure, the raw, the unaltered or unpolished, the organic and untamed. We see natural elements such as earth, water, air, wind, and fire promoted on numerous products, all in order to appeal to natural authenticity. These products may also be experienced as authentic if they are originals.

Products that possess originality in their form, function, or brand are experienced as more authentic offerings than copies, rip offs, imitations, or "me too" products. Another form of authenticity which affects the product value for the customer is the one that may be experienced if a product refers to some other context, drawing inspiration from human history or tapping into shared memories and longings. This referential authenticity may be evoked from any character, time, or location from a small city to a whole continent. This can easily be achieved through 3D printing customizations. Which customer would not want to hold on to a product uniquely made for them to suit their needs, desires, or personality?

The final form of authenticity accounted for by Gilmore and Pine is influential authenticity. According to them, customers also perceive as authentic that which exerts influence on other entities, calling on higher goals and aspirations of a cleaner planet or a better way to live. Merely providing objective value does not suffice for all customers. Some ask themselves the question "How does this change or otherwise influence me, or others, for the better"? Unique customized 3D printed products, for example, made in degradable plastics may, here, contribute with their potential during temporary environmental festivals and other events.

So how can designers improve desirability, authenticity, increase pleasure, and deepen product attachment to extend product life and thus improve product sustainability? There are at least two current design factors that may have a negative effect on design quality and thus product longevity.

The first is manufacturing-design compromise. Because of the restrictive ways in which products currently need to be manufactured, a designers' original design vision has to be compromised to the extent that the product can be made. This means that the product may, potentially, lose some of the desirability the originally envisioned design may have had (Hernandez and Mistree 2000). If, for example, a designer envisages a product component with many overhanging edges, or a very complex form, it may not be possible to injection mold efficiently, and the designer must therefore compromise his vision to the point where the part can be made.

The other factor is that, because of current manufacturing technologies, products are mass-manufactured as one-size-fits-all products which, because of their generic nature, are compromised so as to be useable by all customers but ideal for none (Hunt 2006).

Additive manufacturing is a relatively recent form of manufacturing with the potential to address both these factors, and thus has great potential as an effective tool for more sustainable product design.

3 Additive Manufacturing

The Society of Manufacturing Engineers defines AM as the process of manufacturing a physical object through the layer by layer selective fusion, sintering, or polymerization of a material (SME 2010).

The AM process begins by using a computer generated file to slice a 3D model into thin slices (commonly ranging from 0.01 to 0.25 mm per slice depending on the technology used). The AM machine then builds the model one slice at a time, with each subsequent slice being built directly on the previous one. As a result of the material deposition and processing operations, the digital electronic model is converted into a physical part or product.

Several different AM technologies exist, which differ mainly in terms of the materials they use to build the part, which are typically in a solid, powder or liquid raw state, and the process used for creating the model slices. Until recently, many of these technologies, such as stereolithography (SLA), Fused Deposition Modeling (FDM), early Selective Laser Sintering (SLS), and 3D printing, were only able to make parts for prototyping purposes, as the processes produced parts that were not as strong as injection molded plastic or cast metal parts (Hopkinson et al. 2006).

The latest AM technologies, however, now allow full-strength polymer and metal parts to be produced within hours rather than days (Wohlers 2015). The main technologies that can, today, be classified as rapid manufacturing technologies (as opposed to rapid prototyping) are SLS, Selective Laser Melting (SLM), and Electron Beam Melting (EBM). These technologies create the part by spreading a very thin (typically less than 0.1 mm) layer of powdered material, and then selectively fusing the powder for the appropriate parts of the digital slice of the model. Another layer of powder is then spread on top of the previous one and that is again selectively fused for that slice of the model, at the same time fusing it to the layer beneath. SLS/SLM uses a laser beam for the fusing operation, whereas EBM uses an electron beam to melt the material. The unmelted powder acts as a support material for all the layers above it. It is important to note that, although some of the technologies that use material in solid form for extrusion and those that use photosensitive liquid polymer cured via UV are, in some cases, usable for real manufacturing applications, the vast majority of current manufacturing applications of AM use material in powder form.

Unlike subtractive manufacturing, where material is removed from a larger block of material until the final product is achieved, most AM processes do not yield excessive waste material. As the part is made from material in a powder or liquid form, whatever powder or liquid does not get hardened by the process gets used for subsequent parts. AM typically also does not require the large amounts of time needed to remove unwanted material, consequently reducing time and costs, and producing very little waste (Wohlers 2015).

It is only in the last few years that AM has been used by more and more companies as a viable production technology. As new polymer and metal materials are developed and the speed and precision of the machines further increase, more AM machines are likely to find their way into mainstream production lines (Wohlers 2015).

3.1 Complexity for Free

Additive manufacturing enables the creation of parts and products with complex features, which could not easily have been produced via subtractive or other traditional manufacturing processes. Injection molded or die-cast parts, for example, must be removable from the die in which they are made and must therefore be designed in such a way that this can be done. The metal part shown in Fig. 2, for example, could not easily be machined or cast because there is no way of removing the internal part of the die from the component or of machining the interior

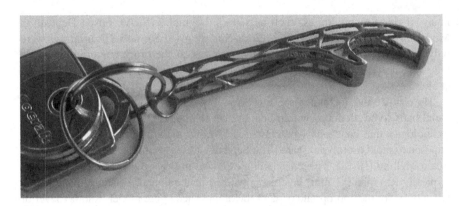

Fig. 2 Additively manufactured titanium bottle opener in which a 90 % material saving was made compared to a bottle-opener made though CNC machining, as conventional manufacturing would be unable to achieve the internal complexity of the component. It was produced in titanium 64 on a selective laser melting system using topology optimization to remove as much material as possible while maintaining the required mechanical characteristics (Diegel 2015)

surfaces. AM, however, does not suffer from these particular restrictions. The complexity of the part does not affect whether it can be made, or even its cost. It allows for components of almost any complexity, freedom in design, and increased flexibility in the features and functions of the end product.

With AM it is also possible to manufacture complex interlocked moving parts in ready-made working assemblies. Though two components may be permanently linked together, they are made as a single component and come out of the machine assembled and ready to work. This is possible because, with the laser sintering process, only the material that forms the component is melted by the laser. The powder that fills the gaps between the moving parts is not melted by the laser so that, once the part is finished, the loose powder between the moving parts is blown out and we are left with a moving assembly. Figure 3 shows a foldable guitar stand

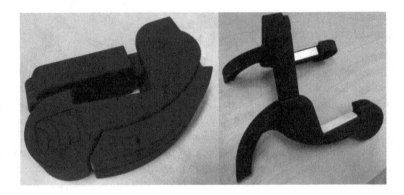

Fig. 3 Folding compact guitar stand, manufactured as a single laser sintered nylon component with integral moving parts (Diegel 2015)

made on a SLS system out of polyamide material (nylon), which is composed of eight different components that allow the stand to unfold and operate in the correct way. The entire guitar stand is, however, manufactured in a single operation with no assembly whatsoever required. If the stand were to be manufactured using traditional manufacturing methods, it would require at least eight components and an assembly procedure to attach all the separate components together.

3.2 Mass Customization

With AM, parts can be immediately made as there is no longer a long lead time to get tooling produced. This has a great impact on new product time to market, and on the ability to produce model changes easily throughout the life of a product. It also has implications in stock control: As components can be made on the spot, companies may no longer need to hold a stock of spare parts as they simply manufacture the parts when needed.

From a product design perspective it also means that every component made can be completely different to the others in a production run without significantly affecting the manufacturing cost. This opens the door to mass customization in which, though mass-manufactured, each product can be customized to each individual customer. Pine and Gilmore (2000) present a framework of mass customization based on four approaches, Transparent, Collaborative, Adaptive, and Cosmetic Customization. For example, with Adaptive Customization 'standard goods and services can be easily tailored, modified, or reconfigured to suit each customer's needs' (Pine and Gilmore 2000). When a patient orders a new product, their personalized data is acquired and is used to modify the basic design configuration to match their data perfectly. The customized components are then fabricated through AM, and the customer ends up with a product customized to them.

The range of personalized data is, of course, enormous and can range from specific shape and size data to full digitized body scans and even MRI scans for internal organs or bones. Some excellent tools to acquire this data already exist (such as laser scanners, body scanners, MRI machines, etc.), and more are being developed as data acquisition technologies improve. Then there is all the personal taste data such as color, texture, mode of use, and more that all need to be acquired in order to help further with product customization.

For this new way of designing products to be used effectively, the product design and the computer aided design industries need to develop new methods for integrating personalized customer data into their designs. This development has already started, particularly in the hearing aid and the dental industries, in which specialized software exists to automate the processes from patient data acquisition to part production. This now needs to be extended to encompass others, including consumer product industries. An example of this is shown in Fig. 4, in which a golf putter was produced on which the handle was customized to fit the player's hands perfectly.

This ability to customize every product made has the potential to affect greatly the desirability, and therefore the longevity, of those products. Are customers not

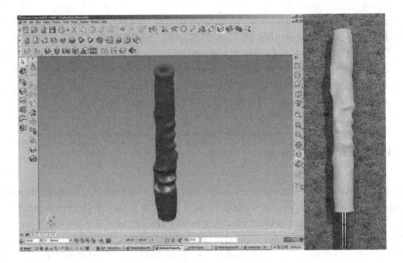

Fig. 4 Custom golf putter handle produced by laser scanning a clay handle that was perfectly formed to fit the user's hand grip, and then additively manufactured in ABS using a fused deposition modelling AM system (Diegel 2015)

more likely to cherish and keep a product that has been specially customized to their particular needs? Though there is little quantifiable data to answer this question as mass customization is still an emerging field, anecdotal data from the high-value custom-made products would seem to indicate that it is, indeed, the case (Mugge et al. 2004).

3.3 Freedom of Design

Because of traditional manufacturing technology restrictions, a product which the designer may have originally envisioned as having certain aesthetics and functionality may need to be compromised so that it can be cost-effectively made. Most designers are quite accustomed to hearing the response of "it cannot be made like that" from manufacturing engineers. They may then need to compromise their design to the extent that the product loses the essence that truly embodies the designers' vision. If this becomes the case then one must ask if the product thereby becomes less desirable and therefore loses some of the longevity it may have had had it been able to be manufactured to the designer's original vision?

With AM, complexity and geometry no longer affect manufacturability. Almost anything the designer imagines can be made precisely as the designer conceived it (Fig. 5).

If one accepts that design practitioners, through their roles in shaping the future, are viewed as being able to promote change in society, especially around unsustainable behaviors (Sosa and Gero 2008), then one must conclude that the product designer is the person with the best ability to create true objects of desire.

Fig. 5 Spider 3D printed guitar by ODD Guitars. These products are manufactured using selective laser sintering and could not have been cost-effectively made using traditional manufacturing methods (Diegel 2015)

If that is the case, then giving them a tool such as AM which allows them to materialize their vision is an absolute necessity. It bypasses the common problem of the design team being told by the manufacturing team that what they envision cannot be made.

It should be noted that AM does not remove all manufacturing restrictions. It replaces them instead with a different set of design considerations which designers must take into account if they wish to use the technologies successfully. These new design considerations are, however, much easier for designers to both understand and comply with without them affecting design intent in a major way. As AM evolves, an entire new 'Design for Additive Manufacture' methodology needs to be developed to maximize the potential the technologies have to offer. Some work in this area has already begun with some researchers at Loughborough University proposing an online design for an additive fabrication feature repository which helps designers to use the correct technologies with the correct feature designs (Campbell 2008).

3.4 Sustainability of Additive Manufacturing Beyond Design Freedom

Beyond the effect AM can have on product design, such as the freedom of design described above, which can be somewhat difficult to quantify, the introduction of AM technologies through the lens of sustainability raises the question of how AM can positively contribute to sustainable development in comparison to traditional manufacturing technologies. A comparative view of the quantified resource consumption (e.g., raw materials and energy usage) of AM and traditional techniques can address such questions (Kianian and Larsson 2015). AM has certain sustainability advantages that are easier to quantify. Chief amongst these is material usage. With metal AM, the vast majority of unmelted powder in the AM process can be directly reused, after a simple sieving operation. The process produces very little material waste in comparison to conventional manufacturing. When machining a component through a traditional subtractive manufacturing process, one starts with a block of metal and machines away a large proportion of that metal to produce the final component. All the metal that is machined away must be either recycled or disposed of.

Similarly, with polymer AM technologies such as SLS, most machines run on a blend of new and used powder, typically using a blend of 50 % new powder with 50 % used powder. So, if the build platform is efficiently filled with nested parts so as to produce a roughly equal volume of sintered and unsintered powder, the system produces very little material waste. In an efficient operation the waste from an AM system can be less than that of injection molding (IM), on which the sprues, runners, and gates that lead molten polymer to the mold cavities must be recycled or disposed of.

From an energy usage point of view, if one examines only low-volume (and high-value) products, AM systems do not differ greatly from conventional manufacturing systems so are, roughly, comparable. One of the initial studies into the relationship of AM to energy consumption was authored by Luo et al. (1999), who found energy measurement results from three forms of AM technologies, namely, stereolithography (SL), laser sintering (LS), and fused deposit modeling (FDM). Later studies focused on contrasting AM to other manufacturing techniques. One such study showed that, during low production volumes, AM had a lesser energy consumption that IM. The energy-based crossover production volume (the position when the compared technologies utilize an equal amount of energy) spans hundred to thousands, depending on part size and geometry (Telenko and Seepersad 2011). When you specifically look at AM production costings, it was first assumed that AM would demonstrate a uniform cost distribution which was unaffected by the volumes of product produced. It was shown that the estimated economic crossover volume (the position when the compared technologies have an equal cost per unit produced) of AM and IM was 14,000 units (Hopkinson and Dickens 2003).

Another study by Ruffo et al. (2006) utilized full costing and the average cost per part produced via AM increased and the corresponding cross-over position decreased to 9000 units. The same author also researched the distinctive capacity of AM to create mixed products during single manufacturing batches and found that the overhead cost could be decreased through mixing different parts, thereby resulting in an overall cost decrease (Ruffo et al. 2006). If, however, one is looking at large-volume mass-manufactured products, then conventional manufacturing technologies are currently much more energy efficient than AM, mainly because of the very short cycle-time of each component in comparison to the relatively slow production speed of AM. It should be noted, however, that it is difficult to compare such numbers with precision as it depends on many of the pre-processing and post-processing steps one includes in the process. Does one, for example, include the energy usage of machining an IM tool, or does one just consider the energy usage of the molding process itself? Does one include the energy usage of having to heat-treat a metal AM part through hot isostatic pressing (a commonly used heat-treatment for metal AM parts, but not necessarily used on all parts), or does one just consider the energy usage of the laser melting process itself? There is much research going on around the world to quantify better the energy usage of both AM technologies and conventional technologies, and it can be predicted that, over the upcoming years, we should get much better quantification of these numbers.

It has already been mentioned above that rapid manufacturing could help postpone a product's end-of-life by creating attachment through product personalization and through larger design freedom. AM can also contribute postponing product replacement by allowing repairs of older products in a relatively cheap and efficient manner. For many consumers (but the same is true for companies), cost is an essential factor for product replacement. For a product that is out of order, it makes senses for most people to get rid of a product if the cost of repair exceeds that of replacement (Page 2014). The same applies if the product is still working but no longer effective because the components are simply getting old. With AM technologies it can be quite easy to manufacture a spare part, especially if the manufacturer agrees to supply the specifications available to product owners or to retailers. Apart from specific sectors where spare parts constitute an important part of the companies' business model, such as the automotive industry, for many manufacturers spare parts involve a cost: inventory costs and shipping costs can be high, molds or tooling for the spare part must be preserved, etc. It can be a win–win both for the manufacturer and the customer to let the part be produced in AM facilities.

An EU-sponsored research project, under the seventh Framework Program, Directspare examined the potential of on-demand production to decrease up-front investment indirectly of materials and storage of spare parts (CORDIS 2009). The attractive aspect of this study was the fact that it explored the potential of AM to provide spare parts for existing parts which were manufactured earlier with other technologies, thus pointing towards another use of AM. It also shows that AM could increase the life of existing products without storage issues and manufacturing equipment associated with large volumes of spare parts.

An example of how AM can be used in helping to extend product life for products with which the customer has formed a certain level of attachment is in the case study of a Volkswagen campervan (Fig. 6) with a damaged tent. In this study, the customer had formed a deep attachment to the product because of its 'retro' feel, and when the zippers on the tent section of the campervan broke he searched for a way of repairing them in order to not have to dispose of the product.

The customer produced a CAD model of the broken component and used AM to very quickly test that the component worked. Not only did the AM component

Fig. 6 1960 Volkswagen Camper with canvas tent accessory (lisbethfalling.com)

work but also the customer saw an opportunity to improve the design, so with a few quick iterations of CAD and AM he produced a design that not only allowed him to greatly extend the life of his cherished product but also allowed him to improve its functionality (Figs. 7 and 8).

Finally, the materials used in AM process technologies can be, in most cases, recycled following the same circuit as the traditional manufacturing technologies.

3.5 Speculations on the Impact of Additive Manufacturing

Our current economic models, largely driven by the advent of factories through the first industrial revolution, are based around the mass production of products through a factory environment. In these models, products are manufactured in a factory which could be located anywhere, finished goods are transported to retail stores (often via distribution centers) that hold a number of the products in stock, and the product is, eventually, bought by the customer. In these models, a large part of the cost of goods sold is not in the direct manufacturing of the products, but indirect costs, such as transport, middle-man infrastructure and margins, etc. (Figure 9). Though figures vary hugely depending on the type of product, it is estimated that the manufacturing cost of a product is typically between 10 and 25 % of the retail price of the product (Reimer 2006; Rodrigue 2015).

In manufacturing, as AM technologies improve and material costs eventually get near those of traditional manufacturing methods, AM could entirely replace traditional manufacturing, thus heralding true on-demand manufacturing, and a world in which product customization is the norm rather than the exception. Not only should the manufacturing be on-demand, but it could well become 'home manufacturing' in which each of us has a 3D printing system at home which can

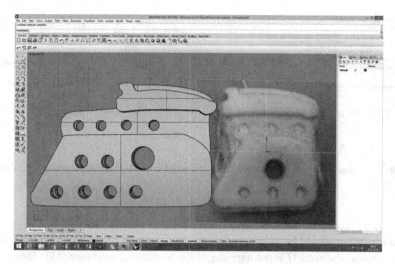

Fig. 7 Improved CAD design next to original broken component (Diegel 2015)

Fig. 8 Selective laser sintering AM produced replacement parts and original broken part (Diegel 2015)

supply whatever manufactured products we need. Your local garage should no longer need to have a storeroom full of spare parts, as they can simply print them as and when needed.

Now extend this thought to our economic system as a whole. Imagine the whole new business systems this could create. Business models in which manufacturing, labor, and transport are no longer part of the equation. Instead, the value

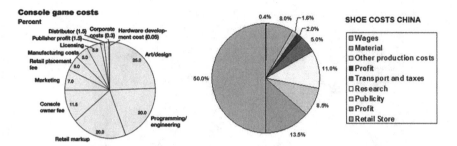

Fig. 9 Examples of the factors that make up the retail cost of a product it (Arstechnica.com; Hofstra.edu)

of products would be in their design, and in the knowledge needed to design and use them. What would be the impact on the environment of not having to transport goods around the world? What happens to the traditional workforce? Do they shift into building AM machines, or servicing, or design?

We are starting to see other areas of business being impacted upon by AM. These include such areas as intellectual property and product liability. On the intellectual property side, as the value of products shift towards being held in the digital version of the product, mechanisms need to be found to protect the inventor of the intellectual property. Some of the intellectual property issues faced by the music industry over the past two decades have been somewhat resolved through mechanisms such as iTunes, Spotify, etc., and these or similar resources may be adaptable to the protection of 3D models. On the product liability side, it is unclear where future product liability lies. If a designer designs a chair and sells that design as a 3D model, and the buyer then modifies the design to suit their needs and then sells it on to a third party who has it manufactured at a local AM service bureau, and the chair breaks, who is liable? The original designer? The person who modified it? The person who made it? This then leads to the areas of AM part qualification and certification which are also areas that need to be resolved over the next few years. How can one guarantee that an AM produced part meets whatever quality standards it is supposed to?

On the medical side, the gradual miniaturization of AM processes allows nanomachines to be printed that can patrol our bloodstream to keep diseases and viruses at bay, and advances in tissue engineering should allow us to print tailored replacement organs. Should we engineer ourselves a spare body to keep on ice, just in case? The impact of such advances and the ethical questions they raise could be tremendous. Beyond the ethical questions, what about the increased leisure time that increased life span is bound to cause?

Even in food, one can imagine one day walking up to our refrigerator and ordering a cheese burger, medium rare, and the fridge printing the requested food for us out of raw materials, coloring, flavoring agents, and all the vitamins necessary to make it as healthy as it can be. Does this mean that conventional cooking now becomes something that is done purely as a leisure activity? And even within this question, are the ingredients as used in leisure cooking grown the old-fashioned way, or are they simply printed?

There are a great many such questions that could arise over the next few years as new AM technologies are developed in different areas of industry. In fact, it could be surmised that the engineering development of AM technologies is possibly ahead of their social, ethical, and business implications.

3.6 New Sustainability Challenges Associated with Additive Manufacturing

With new technologies come new challenges. Just as AM solves some sustainability problems, it can bring new sustainability issues. One of these is the energy issue. Most researchers have agreed that AM is economically costly and energy intensive with high production volumes (Telenko and Seepersad 2011; Hopkinson and Dickens 2003; Ruffo and Hague 2007). Though at lower production volumes, AM derives an advantage because of its decreased up-front investment, specifically in comparison to IM, which has high investment cost with respect to tooling (Telenko and Seepersad 2011). A comparative study of SLS and IM energy consumption presented the energy-based crossover volume to be as low as 50 units for 130-mm products when contrasted with IM using recycled steel molds. This showed the production volume whereby the contrasted manufacturing method had the same energy consumption per unit produced, displaying the energy consumption per unit produced as lower than the other alternative manufacturing method at lower volumes. When the production volume traverses the crossover position, the energy consumption per unit produced flips for the two methods. During this study the crossover volume was shown to increase to around 300 units when contrasted to IM using either virgin steel molds or recycled aluminum molds (Telenko and Seepersad 2011). The energy-based crossover production volume rose to 1500 and 3200, when part sizes were decreased to 45 mm, compared to IM with virgin steel molds and 20 % recycled aluminum molds, respectively. The same study also looked at the cost and found that the economic crossover volume differed greatly depending on which factors were under consideration. They observed that a 35-mm AM manufactured product is estimated to have a crossover volume of 14,000 units when contrasted with IM. When operational overhead costs were included this figure dropped to 9000 units (Hopkinson and Dickens 2003).

If mass customization increases attachment and extends the product life, it creates other issues. With mass production, packaging and logistics are optimized. The packages can have minimal surfaces given a certain level of protection and large quantities can be shipped with low carbon footprints (for many products, packaging and distribution are negligible in terms of environmental impact). With product personalization, it might be necessary to have over-dimensioned packaging. Mass customization also enables the design and manufacture of products in smaller quantities: the shipping of smaller batches or even individual parcels might significantly increase the environmental impact.

AM allows a more local production, which should eventually decrease some of these problems. However, trust must be established between the company that develops the product and the AM manufacturing companies; the product quality must be assured, the companies must agree on time-to-order, etc. Establishing supplier relationships takes time. In the mass production paradigm, companies have a restrained numbers of competing suppliers and this might be the case for AM as well. Coming back to local production, it must be said that many products could be hybrid: parts manufactured with AM and other parts with traditional systems. For these products, the "localization" allowed by AM could be substantially reduced. New logistic models must be worked out.

Products for which the designer has more freedom are often premium products. This can have the side effect of having products use more material than mass production products. One of the reasons is that many of these products are premium products and companies are more driven by adding value to their products than by reducing their costs. In mass production, reducing material quantity by, say, 100 g of steel in 2 million units reduces material costs by $300,000. Such issues cease to be apparent, at least to a certain degree, with personalization. This constraint also disappears from the brief of a designer if he or she is to design an attractive product with more freedom. It therefore becomes important to have a design for AM become a core component of every design process.

The number of types of materials available for AM is much lower than that of materials available for classical manufacturing systems. This is currently a limitation for the design of complex and specific parts. Plastics and steel suppliers can almost develop a specific material on demand for subtractive manufacturing systems but need to ramp up for AM.

Finally, some of the advances in manufacturing that AM makes possible might have both positive and negative environmental impact. One example is possible thanks to the additive principle to design product with heterogeneous materials. An example by Fadel (2004) is a flywheel developed using effectively multi-material. The same flywheel would have been heavier if just the less performant material had been used. In other words, expensive materials (raw material price is often correlated to scarcity) are used only where required by the structural constraints. However, recycling raises another issue.

The real issue behind many of these challenges is not the AM technology per se but the difficulty in determining the environmental consequences of choosing AM over SM. This is a question of having the right environmental assessment and design tools to make the best educated guess.

4 Conclusions

Sustainable product design is about creating products which, although maximizing their economic and social impacts, minimize any harmful effects they may have on the environment. One design philosophy, which can help to achieve this, is to strive towards designing products that become lasting objects of desire, and have a deep attachment between product and user. Products that meet these criteria generally have a greatly increased life span, and this increased longevity reduces the products' negative impact on the environment.

Additive manufacturing, because it allows designers almost unlimited freedom of design, and allows for mass customization of consumer goods, offers the potential for creating such lasting objects of desire, pleasure, and attachment. AM is already beginning to be used in high-value medical products such as hearing aids, medical implants, and the aviation, automotive, and marine industries.

To use AM to its maximum potential, designers need to develop an appropriate set of design methodologies and rules both to incorporate the new features it allows and to take into account the new set of manufacturing restrictions it imposes.

Existing tools, such as LCA, and design frameworks, such as cradle to cradle, need to be adapted to fit the new paradigms of on-demand manufacturing and find ways of being applied earlier in the design process. Likewise, some of the frameworks about what constitute sustainability may need to be revised to reflect better the possibilities of advanced manufacturing technologies.

As AM technologies continue to progress from rapid prototyping to manufacturing, more new materials become available, and multiple material technologies are developed to the point where complex multi-material production quality assemblies can be made, the field of product design needs to evolve in parallel so as to meet better the demands of emerging sustainable design trends.

References

Abukhader SM (2008) Eco-efficiency in the era of electronic commerce—Should an eco-effectiveness approach be adopted? J Clean Prod 16:801–808

Accenture (2014) Circular advantage: innovative business models and technologies to create value in a world without limits to growth

Arstechnica.com, Costing on production video games, http://arstechnica.com/gaming/news/2006/12/8479.ars. Accessed Nov 2015

Braungart M, McDonough W, Bollinger A (2007) Cradle to cradle design: creating healthy emission—A strategy for eco-effective product and system design. J Clean Prod 15:1337–1348

Bulow J (1986) An Economic Theory of Planned Obsolescence, Quarterly Journal of Economics, 101:729–749

Campbell RI (2008) Creating a design for additive fabrication feature repository, RAPDASA 2008, Rapid Product Development Association of South Africa Conference, South Africa

Ciambrone DF (1997) Environmental life cycle analysis. CRC Press LLC, Florida

CORDIS: Community Research and Development Information Service (2009) Strengthening the industries competitive position by the development of a logistical and technological system for spare parts that is based on on-demand production (DIRECTSPARE)

Datschefski E (2004) The Total Beauty of Sustainable Products, London, BioThinking Internatio nal,http://www.biothinking.com/btintro.htm. Accessed January 2016

Diegel O (2015) Figures and diagrams for this book chapter

Dreyer L, Niemann A, Hauschild M (2003) Comparison of three different LCIA methods: EDIP97, CML2001 and eco-indicator 99. Int J Life Cycle Assess 8(4):191–200. doi:10.100 7/BF02978471

Fadel G (2004) Rapid prototyping and manufacturing technologies: accomplishments and potentials. 5th international symposium on tools and methods of competitive engineering—TMCE'04, vol 1, Lausanne, 13–17 April 2004, pp 29–47

Fuad-Luke A (2009) The eco-design handbook: a complete sourcebook for the home and office, 3rd edn. Thames & Hudson, London

Gilmore JH, Pine BJ (2007) Authenticity: what consumers really want. Harvard Business School Press, Boston, pp 45–79

Glavic P, Lukman R (2007) Review of sustainability terms and their definitions. J Clean Prod 15:1875–1885

Goedkoop M, Effting S, Collignon M (2000) The eco-indicator 99: a damage oriented method for life-cycle impact assessment: manual for designers. PRé Consultants B.V., Amersfoort

Govers PCM, Mugge R (2004) 'I love my Jeep, because it's tough like me': the effect of product-personality congruence on product attachment, 2004 international conference on design and emotion, Ankara, Turkey

Halweil B, Mastny L, Assadourian E (2004) State of the world 2004—Special focus: The Consumer Society. Worldwatch Institute, Washington DC

Hernandez G, Mistree F (2000) Integrating product design and manufacturing: a game theoretic approach. Eng Optim 32(6):749–775

Hofstra.edu, Graphs on show production costs in China, http://people.hofstra.edu/geotrans/eng/ch5en/appl5en/costs_shoe_China.html. Accessed Nov 2015

Hopkinson N, Dickens PM (2003) Analysis of rapid manufacturing-using layer manufacturing processes for production. Proc Inst Mech Eng Part C J Mech Eng Sci 217:31–39

Hopkinson N, Hague RJM, Dickens PM (2006) Rapid manufacturing an industrial revolution for the digital age. Wiley, New York

Hunt D (2006) PhD thesis: a consumer perspective on mass customization, University of Missouri-Columbia, May 2006. Available from http://edt.missouri.edu/Winter2006/Dissertation/HuntD-052506-D4001/research.pdf. Accessed Feb 2010

Kemp R, Martens P (2007) Sustainable development: how to manage something that is subjective and never can be achieved? Sustain Sci Pract Policy 3(2):5–14

Kianian B, Larsson TC (2015) Additive manufacturing technology potential: a cleaner manufacturing alternative. 20th design for manufacturing and the life cycle conference (DFMLC). Proceedings of the ASME 2015 international design engineering technical conferences & computers and information in engineering conference (IDETC/CIE 2015), 2–5 Aug 2015, Boston, MA, USA

Kobayashi H (2006) A systematic approach to eco-innovative product design based on life cycle planning. Adv Eng Inform 20:113–125

Le Pochat S, Bertoluci G, Froelich D (2007) Integrating ecodesign by conducting changes in SMEs. J Clean Prod 15(7):671–680. doi:10.1016/j.jclepro.2006.01.004

Ljungberg LY (2007) Materials selection and design for sustainable products. Mater Des 28:466–479

Luo Y, Ji Z, Ming L, Caudill R (1999) Environmental performance analysis of solid freeform fabrication processes. In: Proceedings of the 1999 IEEE international symposium on electronics and the environment, 11–13 May 1999

McDonough W, Braungart M (2002) Cradle to cradle: remaking the way we make things. North Point Press, New York

McDonough W, Braungart M, Anastas PT, Zimmerman JB (2003) Applying the principles of green engineering to cradle-to-cradle design. Environ Sci Technol 1:434–441

Missimer M (2015) Social sustainability within the framework for strategic sustainable development. Doctoral thesis, Blekinge Institute of Technology, Faculty of Engineering, Department of Strategic Sustainable Development. ISBN 978-91-7295-307-9

Mugge R, Schifferstein HNJ, Schoormans JPL (2004) Personalizing product appearance: the effect on product attachment. 2004 international conference on design and emotion, Ankara, Turkey

Packard V (1978) Waste makers. Simon & Schuster. ISBN 0671822942

Page T (2014) Product attachment and replacement: implications for sustainable design. Int J Sustain Des 2(3):265–282. doi:10.1504/IJSDES.2014.065057

Papanek V (1985) Design for the real world: human ecology and social change, 2nd edn. Academy Chicago Publishers, Illinois

Park M (2005) Sustainable consumption in the consumer electronics sector: design solutions and strategies to minimise product obsolescence. 6th Asia Pacific roundtable for sustainable consumption and production (APRSCP); Melbourne, Australia, Oct 2005

Pine J, Gilmore J (2000) Markets of one: creating customer-unique value through mass customisation. Harvard Business Review Book, Boston

Reimer J (2006) Costing on production video games, http://arstechnica.com/gaming/news/2006/12/8479.ars. Accessed Oct 2015

Rodrigue JP (2015) Graphs on show production costs in China, http://people.hofstra.edu/geotrans/eng/ch5en/appl5en/costs_shoe_China.html. Accessed Oct 2015

Ruffo M, Hague R (2007) Cost estimation for rapid manufacturing- simultaneous production of mixed components using laser sintering. Proc Inst Mech Eng Part B J Eng Manuf 221(11):1585–1591

Ruffo M, Tuck C, Hague R (2006) Cost estimation for rapid manufacturing-laser sintering production for low–medium volumes. Part B J Eng Manuf 220:1417–1428

Sherwin C (2004) Design and sustainability: a discussion paper based on personal experience and observations. J Sustain Prod Des 4:21–31

SME (2010) Available from http://www.sme.org/cgi-bin/communities.pl?/communities/techgroups/ddm/what_is_ddm.htm&&&SME. Accessed Feb 2010

Sosa R, Gero JS (2008) Social structures that promote change in a complex world: the complementary roles of strangers and acquaintances in innovation. Futures 40:577–585

Telenko C, Seepersad CC (2011) A comparative evaluation of energy consumption of selective laser sintering and injection molding of nylon parts. In: Proceedings of the 2011 solid freeform fabrication symposium, pp 41–54

Telenko C, Seepersad CC, Webber ME (2008) A compilation of design for environment principles and guidelines. 13th design for manufacturing and the lifecycle conference—DETC/DFMLC'08, vol 5, New York, NY, 3–6 Aug 2008, pp 289–301. doi: 10.1115/DETC2008-49651

Tischner U, Charter M (2001) Sustainable product design. In: Charter M, Tischner U (eds) Sustainable solutions: developing products and services for the future. Greenleaf Publishing Limited, Sheffield, pp 118–138

Vallet F, Eynard B, Millet D, Glatard MS, Tyl B, Bertoluci G (2013) Using eco-design tools: an overview of experts' practices. Des Stud 34(3):345–377. doi:10.1016/j.destud.2012.10.001

Van Nes N, Cramer J (2003) Design strategies for the lifetime optimisation of products. J Sustain Prod Des 3:101–107. Springer, Berlin 2006

Van Nes N, Cramer J (2005) Influencing product lifetime through product design. Bus Strategy Environ 14(5):286–299

Vincent J (2006) Emotional attachment and mobile phones. Knowl Technol Policy 19(1):39–44

Waage SA (2007) Re-considering product design: a practical "road-map" for integration of sustainability issues, Journal of Cleaner Production, Vol.15, No.7, pp. 638–649

Waldman M (1993) A New Perspective of Planned Obsolescence, The Quarterly Journal of Economics, pp. 273–283.

Wenzel H, Hauschild M, Alting L (1997) Environmental assessment of products, vol 1. Chapman and Hall, London

Whiteley N (1993) Design for society. Reaktion Books, London

Wohlers T (2015) Worldwide progress report on the rapid prototyping, tooling, and manufacturing state of the industry. Wohlers report 2009, Wohlers Associates, USA

Zafarmand SJ, Sugiyama K, Watanabe M (2003) Aesthetic and sustainability: the aesthetic attributes promoting product sustainability. J Sustain Prod Des 3:173–186

Sustainable Design for Additive Manufacturing Through Functionality Integration and Part Consolidation

Yunlong Tang, Sheng Yang and Yaoyao Fiona Zhao

Abstract Additive Manufacturing (AM) is a type of material joining process whereby parts can be directly fabricated from its 3D model by adding materials typically in a layer by layer fashion. Compared to conventional manufacturing techniques, AM has some unique capabilities which bring significant design freedom for designers. Some of this design freedom is manifested in the innovative design of lattice structure to achieve multifunction with reduced weight and consolidated component designed with reduced part count and improved performances. A new type of design philosophy for AM is emerging that is to achieve integrated functions and part consolidation, which plays a significant role in sustainable design. This chapter discusses this new design philosophy with a thorough review of lattice structure design and optimization methods, design for AM methods, and other related new design methods. It presents a general design framework to support sustainable design for AM via functionality integration and part consolidation. This proposed general design methodology supports the design that has less part counts and less material but without compromising its functionality. A case study is given at the end of the chapter to illustrate and validate the proposed design methodology. The result of this case study shows that the environmental impact of a product's manufacturing process can be reduced by redesigning the existing product based on the proposed design methodology. Moreover, compared to its original design, the redesigned product also has a lower part count. Generally, this case study implies that design freedom enabled by AM is an indispensable factor which needs to be considered during the environmental impact analysis of products fabricated by AM processes.

Keywords Additive manufacturing · Environmental impacts · Functionality integration · Life cycle assessment · Parts consolidation · ReCiPe midpoints indicator

Y. Tang · S. Yang · Y.F. Zhao (✉)
Department of Mechanical Engineering, McGill University, Montreal, QC, Canada
e-mail: yaoyao.zhao@mcgill.ca

© Springer Science+Business Media Singapore 2016
S.S. Muthu and M.M. Savalani (eds.), *Handbook of Sustainability
in Additive Manufacturing*, Environmental Footprints and Eco-design
of Products and Processes, DOI 10.1007/978-981-10-0549-7_6

1 Introduction

Additive Manufacturing (AM) is a new emerging technology that joints material typically in a layer by layer fashion. It has been increasingly used in the new product developing spanning conceptual design, functional design, and tooling. AM is also referred to as 3D printing, rapid prototyping, solid freeform fabrication, and direct manufacturing. It has shown great promise in applications to medical implants and the aerospace and automobile industries (Hopkinson et al. 2006; Gibson et al. 2010; Murr et al. 2010). A variety of raw materials can be used in AM processes including metal, plastic, ceramic, sand, and composites. AM can build a part directly from a digital representation without tooling and fixtures. Although AM is inherently suited for making products with high complexity in small batches, it has shown great capability to accelerate mass production by making tools and dies used in large volume manufacturing. It can also accelerate the production of selected parts by combining multiple parts into one.

The AM fabrication process has many unique characteristics that are very different from conventional manufacturing and proper selection of the process parameters significantly affects the product's quality. Thus, planning decisions to select an appropriate AM process and its parameters for specific application requirements and design are rather involved. Extensive research work has been done to analyze the influence of AM process parameters on end product quality such as surface finish, dimensional accuracy, and mechanical properties (Singhal et al. 2009; Byun and Lee 2006; Delgado and Ciurana 2012). With proper fabrication parameters, functional products with high complexity, multi-functions, reduced part count, and high added value can be produced without significant increase of manufacturing cost. Thus, this process is widely described as a "clean" or "green" process because only the exact amount of material to build the functional parts is needed (Bourhis et al. 2014). However, such claim needs to be assessed quantitatively with the consideration of the entire life cycle and only very limited research has been done on this topic. Most reported research in this area has been aimed at:

1. Developing general process models and environmental evaluation methods (Luo et al. 1999; Le Bourhis et al. 2013)
2. Understanding the environmental impact of specific AM processes (Kellens et al. 2011, 2012)
3. Measuring life cycle inventory data of specific AM processes (Sreenivasan et al. 2010; Xu et al. 2014)
4. Comparing the environmental impacts of different AM processes and conventional manufacturing processes (Mani et al. 2014; Morrow et al. 2007; Yoon et al. 2014; Faludi et al. 2015)

Most research was conducted within the established framework of Life Cycle Assessment (LCA), through which quantitative environmental impact data can be obtained based on new unit AM process models, assessment boundaries, and cut-offs. When using this method, a static LCA analysis is performed after a product is manufactured, by which time the environmental impacts have already been generated. LCA tools are typically not integrated with other design analysis and process optimization methods. Most comparisons between the environmental impact of AM processes and conventional manufacturing process have so far been done for the same product design (Mani et al. 2014; Morrow et al. 2007; Yoon et al. 2014; Faludi et al. 2015). The use of the same conventional manufacturing product design as input for an LCA study is to a certain extent misleading because the extensive research on Design for Manufacturing (DFM) over the past several decades has established valid and often standard geometric features for product design to maximize manufacturing efficiency and minimize manufacturing cost. Such DFM methods and guidelines do not apply to AM (Yang and Zhao 2015), thereby making the sustainability analysis results of AM technology look less appealing. Deeper analysis reveals that AM technologies provide designers with unique and extensive freedom to optimize further the design to be more environmental friendly without compromising its functional performance. For example, structural optimization methods can be applied to reduce structural weight significantly, which may decrease energy and material consumption during product fabrication. Recent studies show that the fabrication of the optimized product allows one to "cut material consumption by 75 % and CO_2 emission by 40 %" (www.3Ders.org 2013). Thus, it is unfair to study the sustainability of AM processes based on the same product design. The design freedom of AM needs to be considered in the sustainability analysis and evaluation. Furthermore, the design freedom provided by AM enables mass customization within the industry. Another advantage of AM processes is the ease of embedding unique product features to achieve targeted functionalities. Short design to fabrication turnover is another benefit of AM processes that enables quick redesign to improve a product's functional performance as well as its sustainability. It is known that the design has the most influence on the sustainability of a product in its entire life cycle (Seliger et al. 2011).

Thus, this chapter first gives a thorough review of design methods for AM including lattice structure design and its optimization methods, design for AM methods, and other related new design methods. It then presents a general design framework to support sustainable design for AM via functionality integration and part consolidation. This general framework mainly consists of four stages. In the first stage, initial functional design has been done to determine the physical entities based on the input functional specifications. In the second design stages, AM-enabled design optimization methods can be applied to minimize the environmental impact of products manufacturing process based on the pre-feedback of environmental impact evaluation model. Then the optimized FVs are further refined and divided into several different parts based on assembly requirements for its related parts. Some assembly features can be added. Finally, the sustainability evaluation model can be used to estimate the sustainability of design solutions for a given AM process. Based on the result of evaluation, the optimized design solution and its environmental impact during the manufacturing phase can be output. A case study is presented to validate the proposed design framework.

2 AM Enabled Design Methods

It is widely acknowledged that design for environment or sustainability should satisfy all the function requirements and performance requirements in the first place. To facilitate a general design methodology for sustainability in the context of AM, there is a strong need to understand how to satisfy these function requirements and performance requirements in the design process with a priority of sustainability consideration. With AM evolving from rapid prototyping to rapid manufacturing, this novel technology is being widely applied in industry to fabricate functional parts. However, how to effectively employ the design freedom enabled by AM and comply with inherent manufacturing constraints remains undeveloped. To make a breakthrough, the impact of AM on conventional design theory and methodology (DTM) is briefly summarized in Sect. 2.1. With the awareness of this impact, ongoing research on design for AM draws much attention and is discussed in Sect. 2.2. To finalize how to design for sustainability within the scope of AM, current AM-related design research on sustainability is discussed in Sect. 2.3.

2.1 Impact of AM on Conventional DTM

It is asserted that the most useful and practical theories and methodologies are characterized by mathematic foundation, concrete objectives, or explicit processes (Tomiyama et al. 2009). Therefore, this chapter narrows the scope into analyzing the very DTM which matches these characteristics. According to the classification method of DTM proposed by Tomiyama (Tomiyama 2006) based on the General Design Theory (GDT) (Reich 1995), representative design methodologies such as Axiomatic Design (Suh 1990), Systematic design method (Pahl et al. 2007), Design for X (DFX, referring to DFM, DFA, DFMA, and DFD in this chapter), Adaptable Design (Gu et al. 2004), Characteristics-Properties Modeling (CPM) (Weber 2005), and Contact and Channel Model (C&CM) (Albers et al. 2003) all belong to the second category. This category is called "DTM to enrich attributive and functional information of design solutions." The other two categories are "DTM to generate a design solution" and "DTM to manage design and represent design knowledge." AM exerts an influence on all these three categories. How to generate a design solution is changed by functional complexity because more functions are achievable in a single part by AM. How to manage design and represent design knowledge is affected by the no-tooling and sustainable manufacturing methods. However, most influential is the way AM enriches attributive and functional information of design solutions. The impact on this category is reflected on the design considerations for manufacturing, assembly, and performance. For the limitation of content, more information in this section can be found in the authors' review paper (Yang and Zhao 2015).

2.1.1 Design Considerations for Manufacturing

For any functional product, one of the critical steps in the design process is to check manufacturability. Additive manufacturing as a manufacturing process should also follow this procedure because AM still exerts manufacturing constraints on design. Typical manufacturing constraints could be available materials, geometric limitations [such as minimum wall thickness and minimum clearance (Thomas 2010)], dimensional accuracy (Regenfuss et al. 2007) and surface roughness, support design and removal for some techniques such as Selective Laser Sintering (SLS), low mechanical properties [for example Material Jetting process (Wohlers 2010)], building time for large size components, and material recycling [i.e., FGMs (Watts and Hague 2006)]. For conventional manufacturing processes such as machining, forging, injection molding, and so on, DFM rules and practices have already been well exemplified in Handbook for Product Design for Manufacture (Bralia 1986) and Product Design for Manufacture and Assembly (Boothroyd et al. 2002). In contrast, such design rules for manufacturing consideration have not been established yet for AM. The reason may be attributed to a lack of understanding of the physical principles of powder metallurgy and the diversity of AM processes. DFM requires designers to have a good understanding of the manufacturing constraints imposed by available fabrication methods. In this part, the main goal is to illustrate the incompetence of conventional DFM instead of proposing design for AM rules. The challenges for DFM in AM application are reflected in the following aspects:

1. Layer by layer working mechanism and joining material from CAD model data without tooling. This new working principle totally expands designers' imagination in part design. Unlike the subtractive and formative processes, this additive process can virtually build parts in any shapes.
2. Hybrid manufacturing. Parts could advantageously be designed from the modular and hybrid point of view, whereby parts are seen as 3D puzzles with modules. This kind of hybrid manufacturing method can be divided into two categories. The first is the combination of different AM technologies such as the combination of stereolithography (SL) and direct write (DW) in the area of electronics (Perez and Williams 2013; Lopes et al. 2012). The second is the combination of AM and conventional manufacturing methods such as selected laser melting (SLM) and CNC machining.
3. Complex material composition in a controlled manner. Because materials with AM technologies can be processed at each point or at each layer at a time, the manufacturing of parts with complex material compositions and designed property gradients is enabled.
4. Architecture with hierarchical complexity. The AM process enables the fabrication of architecture design of hierarchical complexity across several orders of magnitude in length scale. There are three typical features in reported research

which are tailored nano/microstructres, textures added to surfaces of parts, and additional cellular materials (materials with voids), including foams, honeycombs, and lattice structures.

5. Repair and remanufacture scenario. The unique process characteristics of AM make it possible to remanufacture and repair with low cost and relative high speed.

According to the above five main challenges on DFM rules, several resultant rules should be considered. The first rule is that when considering hybrid manufacturing, for example, CNC machining and SLM, the DFM rules of CNC machining should automatically be considered, such as tool accessibility. The second rule is that, although AM facilitates multiple material deposition, how to find the material combination and how to avoid stress singularity at the interface is critical. The third rule is that cellular structure sometimes could increase manufacturing difficulty because it is difficult to remove the support structure. The fourth rule is that repair and remanufacture is different from the manufacturing process; in such a case, a new set of rules is necessary.

2.1.2 Design Considerations for Assembly

Most products are comprised of multiple parts, which means that assembly considerations are important. From the conventional DFA aspect, two main considerations are often offered to reduce assembly time, cost, and difficulties: minimize the number of parts and eliminate fasteners. Both considerations are translated directly to fewer assembly operations, which is the primary driver for assembly costs (Boothroyd et al. 2002). Traditionally, assembly's main function is to join components, formless material, and sub-assemblies into a complex product (Andreasen et al. 1983). In contrast with conventional assembly processes, AM enables part consolidation in the place where parts used to be fabricated separately because of manufacturing limitations, material differentiation, or cost. Manufacturing limitations are lessened by AM and AM offers a totally different perspective of joining compared to conventional assembly. The challenges for design considerations for assembly in AM processes are discussed in the following:

1. Integrated assembly and embedded components. Layer by layer or point by point characteristics make it possible to realize integrated assembly and embedded components. Typical applications are classified into two groups: operational mechanisms (Mavroidis et al. 2001) and embedded components. In the operational mechanisms case, even when two or more components must be able to move with respect to one another, AM can build these components fully assembled. In the embedded components case, it is often advantageous to embed components into a part to construct a functional prototype to improve systematic performance. These embedded components include small metal parts (i.e., bolts), electric motors, gears, silicon wafers, printed circuit boards, and strip sensors.

2. A special assembly method. Joining multiple materials together by AM is a feasible assembly method. The use of multiple materials within AM to increase part functionalities has been considered by many researchers in the form of FGM. However, there are many fabrication issues to be addressed in these cases in addition to the dilemma of recycling components fabricated from multiple materials. Functionally Graded Rapid Prototyping (FGRP) is a novel design approach and technological framework enabling the controlled spatial variation of material properties through continuous gradients in functional components (Oxman et al. 2012).

2.1.3 Design Considerations for Performance

With AM eliminating much of the manufacturing constraints and assembly needs, designers are partially free from the constraints of design for manufacture and assembly (DFMA), which means that design for performance (DFP) turns into reality. Traditionally, a product with simple geometry is desirable to avoid sacrificing its function or performance because manufacturing cost and difficulty normally increase as structure complexity increases. However, this rule does not fit AM any more. The manufacturing cost and difficulty when using AM is not overly related to structure complexity.

Performance in this chapter is a general term which embraces functional performance and complementary performance. Functional performance normally refers to performance parameters that are directly related to corresponding functions, e.g., lift coefficient. Complementary performance normally refers to products' service life, e.g., reliability. Typical objectives of DFP could be measurable capacity of a design including force, strength, stiffness, stress, aerodynamic properties, heat dissipation, and biomedical properties. For example, heterogeneous structure may result in better performance such as weight reduction, uniform stress distribution, and better cooling effects. However, in a traditional way, this kind of design concept is to be rejected because of manufacturability considerations. With the aid of AM, heterogeneous structures can be achieved in two levels. The first is the material level. Besides FGMs, another possible way is to mix different cell units of the same material within the same design domain; meanwhile, the drawbacks of computational power requirement and dilemma of recycling of FGMs are avoided (Watts and Hague 2006). The second is at meso or macro structure level. This type of heterogeneousness can be achieved by topology optimization or cellular structures.

2.2 AM-Related Design Method

Realizing the incompetence of conventional DTM in adopting the design freedom enabled by AM, many researchers have started to establish various design rules or guidelines to help successfully employ AM in building functional parts. Basically, these approaches can be divided into three categories: design guidelines, modified DTM, and design for additive manufacturing (DFAM).

2.2.1 Design Guidelines and Design Rules

In this category, the main goal is to establish a set of rules to guide design on the basis of full understanding of manufacturing constraints of various AM processes. Basically, this type of method could be regarded as the extension of DFM with a focus on AM. These rules are generally not quantitative in nature and require a human to interpret and apply to each specific and unique case. Whilst this is much better than just blindly starting each design from scratch. There two main deficits of this kind of method: (1) with only one focus on manufacturability, it does not enhance performance improvement and (2) it requires designers to have much AM knowledge to interpret the rules.

As an example, the design guidelines for rapid manufacturing (RM) given by Becker et al. (Becker et al. 2005) are given as follows:

- Use the advantages that are included in RM processes
- Do not build the same parts designed for conventional manufacturing processes
- Do not consider traditional mechanical design principles
- Reduce the number of parts in the assembly by intelligent integration of functions
- Check whether there are bionic examples to fit your tasks as these can give a hint towards better design solutions
- Feel free to use freeform designs; they are no longer difficult to produce
- Optimize your design towards highest strength and lowest weight
- Use undercut and hollow structures if they are useful
- Do not think about tooling because it is no longer needed

Design guidelines focus on a more general discipline where designers are encouraged to make a better design by taking advantages of AM. In contrast, design rules deal with a more specific aspect of identifying the limitations of AM, serving as design code. Abundant research can be found in this area (Thomas 2010; Adam and Zimmer 2014; Popsecu 2007; Kruf et al. 2001; Kim and Oh 2008; Mahesh et al. 2004; Shellabear 1999). Research on design rules can be divided into two groups: experimental method, for example benchmark study, and systematic method. The former is represented by the research of Daniel (Thomas 2010). In his research, the geometric limitations of SLM were evaluated through a quantitative cyclic experimental methodology. Part orientation, fundamental geometries, and compound design features were explored to generate the design rules for the SLM process. A more effective way to verify design rules is to build a benchmark. In benchmark tests (Kim and Oh 2008; Mahesh et al. 2004; Shellabear 1999), mechanical properties such as tensile and compressive strengths, hardness, impact strength, heat resistance, surface roughness, geometric and dimensional accuracy, manufacturing speed, and material costs were compared for different types of AM process. The latter is represented by Adam and Zimmer (2014). The group was working on a project named "Direct Manufacturing Design Rules 2.0" where function-independent design rules were studied for laser sintering, laser melting, and fused deposition modeling AM processes. Within the suggested research flow, geometric elements are first defined as basic elements, element transitions, and aggregated structures. Then, after studying the attribute value, boundary conditions of these groups, design rules are obtained.

In addition, ASTM released general design guidelines (ASTM Standard 2012) for AM including design opportunities and limitations. Design opportunities cover layer by layer manner, possible sophisticated geometry, varied material or property, and design for functionality. Limitations of adopting AM as the fabrication method can be concluded as economical consideration, production volume, material choices, geometry discretization, building envelop, and post-processing. For the design aspect, geometry consideration, material property consideration, process consideration, product consideration, use consideration, sustainability consideration, communication consideration, and business consideration are all reported.

Design guidelines and design rules provide a feasible way to aid designers to design effectively in applying AM technologies; however, this kind of case study orientated guidelines is only suitable for avoiding the restrictions of conventional design rather than providing how to take full advantage of AM-enabled design freedom. It is important to note that most of the design guidelines emphasize how to take advantage of AM capabilities, whereas the unprecedented limitations are rarely studied.

2.2.2 Modified DTM for AM

Adopting a precise and consistent design methodology to design a product is always suggested (Segonds 2011). Boyard et al. (2014) managed to put forward a modified DFMA methodology to improve the design process of AM related design. This design method consists of five steps: functional specifications, conceptual design, architectural design, detailed design, and implementation. It is characterized by the feature that DFA and DFM work in parallel simultaneously rather than sequentially. This feature is enabled by a modular and modifiable function graph in the conceptual design phase, where each function is represented by a sphere node and these nodes are linked by segments to indicate direct relationships of functions and spatial locations. Once these nodes and links are established, functional sets are determined by the criteria oriented from DFA against which each part should be examined as it is added to the product during assembly (Boothroyd et al. 2002). A function graph of sets was proposed to model a product and each set represents a part, different sets being connected by dotted lines (see Fig. 1). This kind of function graph allows users to recognize functions and functional relationships spatially. However, whether it is reasonable to link function A and function B is not given. For example, function A and function B both belong to set Ω by proposed criteria whereas the relationship between A and B

(a)   (b)

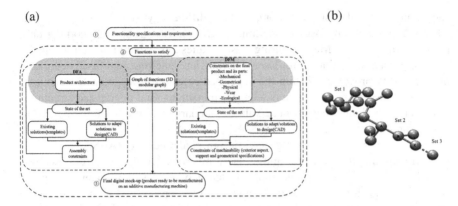

Fig. 1 Design methodology proposed by Boyard and Rivette (Boyard et al. 2014). **a** Modified DFMA method. **b** Function graph

is not defined. This proposed design methodology facilitates the idea of considering DFA and DFM simultaneously in AM design processes although it is not well developed for complete AM design innovation. For instance, it does not deal with a product incorporating inner relative movement and hierarchical complexity.

To develop a design methodology specially for AM, Rodrigue (2011) asserted that DFA and DFM were the only possible design methodologies related to AM. In the case of AM, geometry constraints and assembly difficulties were proven to be less important. To optimize the product with respect to assembly and manufacturing, DFA and DFM were performed to meet the initial user's requirements. Then a redesign methodology was proposed to optimize products for preventing failure and to meet user requirements. Prevention of failure was based on FMECA (Failure Modes, Effect and Criticality Analysis) which was derived from FMEA. It aims to increase the reliability to meet the specifications. Compliance with user requirements aims to meet the design constraints with minimum compromise. Finally, the optimization is examined to decide the structure and shape of the final product. This method concentrates more on design reliability whereas how to meet user requirements are not clearly discussed.

2.2.3 Design for Additive Manufacturing

Design for additive manufacturing (DFAM) could be regarded as the evolvement of design for rapid manufacturing (DFRM) in its early days. However, DFAM in this section is not referring to additive manufacturing as a manufacturing process. Instead, DFAM is focused on how to adopt the design freedom of AM fully to improve product performance. In other words, this section concentrates on design methods. These design methods for AM can be put into two groups. The first concerns AM-enabled structural optimization design methods, the second DFAM methodology.

Generally, structure optimization-related design methods are more specific with concrete objectives. AM-related structure design optimization methods can be classified by different objectives including stiffness, strength, compliance, and stress distribution in static structure design. In addition, structural optimization methods have spread to other disciplines such as dynamic (Evans et al. 2001; Ma et al. 2006), thermal (Zhou et al. 2004; Blouin et al. 2005; Rännar et al. 2007), and biomedical fields (Chen et al. 2011; Castilho et al. 2013; Faur et al. 2013). According to whether the optimization process considers manufacturing constraints, these optimization-related design methods can be grouped into two categories: unconstraint optimization and constraint optimization. In the early stages of design for AM, most researchers focus on the former to explore the potential of functionally optimal geometric design solutions. The means to realize optimal design could be a geometric way including parametric optimization, geometric optimization (shape, size, and topology), and cellular structures, or a material way, for example, functionally graded material. Taking topology optimization as an example, typical topology optimization methods include the ground structure method (Bendsøe et al. 1994; Dorn et al. 1964), homogenization method (Bendsøe and Kikuchi 1988), Solid Isotropic Material with Penalization (SIMP) method (Rozvany et al. 1992), level set method (Allaire et al. 2002; Wang et al. 2003), evolutionary method (Xie and Steven 1993; Young et al. 1999), and genetic method (Wang and Tai 2005; Chen et al. 2009). In contrast to topology optimization, cellular structures such as lattice can be optimized in terms of pattern (uniform or conformal), cell topology, strut thickness, material, orientation, lattice skins, and so forth. It is worth mentioning that it requires domain-specific knowledge to interpret some objectives to establish the relationship between design variables and objective functions for structural optimization problems. For example, to design a scaffold for tissue engineering, the desired structure is supposed to support the proliferation of cells. For constraint optimization method, manufacturability is greatly emphasized in the optimization process. A manufacturability check could be done simultaneously or iteratively with the help of design rules. Constraint optimization design method is becoming predominant in the industry when choosing AM as a new process. An overall DFAM computer aided system framework has been developed by Rosen (2007) consisting of part and specification modeling, process planning, and manufacturing simulation. In this design flow, the emphasis is placed on material and cellular structure modeling and optimization with respect to the manufacturing support module.

In contrast, DFAM design methodologies concentrate on how to design in a more general way without concrete objectives. They cover not only downstream design activities such as parametric optimization and DFM, but also certain upstream design activities such as functional design and part consolidation in the early design stage. Vayre et al. (2012, 2013) proposed a design method consisting of four steps: analyze the specifications, initial shape, parametric optimization, and validation of manufacturability. Manufacturing constraints are dedicated to laser-based or EBM-based AM processes including accessibility constraints, frequent acceleration and deceleration stages, heat dissipation, and inability to build closed hollow volume. This method indicates the need for functional design; however, the method is way too general so that how to generate initial shape and perform parametric optimization is not illustrated. Rosen (2007) and Tang et al. (2014) also indicate the need for satisfying the design specification and consider the extra step of how to conduct structural optimization to achieve multifunctional and multilevel design. In their design methods, multiple functions are explicitly referred to structural performance (i.e., stiffness or stress distribution) and conjugate heat transfer or vibration absorption, and multilevel design is strictly limited to cellular structures. These methods share a common deficit that manufacturability is not considered and the initial design space is given. To incorporate the capability of AM in realizing part consolidation, Yang et al. (2015) proposed a new part consolidation method to integrate function integration with structural optimization to achieve a better performance and lower part count with respect to manufacturing constraints, assembly requirements, and modularization requirements. One of the main contributions is that it proposes a feasible way to deal with assembly design in the context of AM, which is one of the main deficits of most current designs for AM research because assembly is not considered.

Design is always creative work, especially the early design stage. Although how to optimize structure and achieve function integration is becoming known, producing certain creative shapes or structures could be exhausting without a good knowledge of AM. To ease the difficulty of coming up with innovative shapes, a feature-based design approach was proposed by Bin Maidn (2011). These design features serve as an inspiration for designers in the conceptual design stage. However, several issues remain unsolved including the AM feasibility validation approach adopted in the system and how to do morphing on the basis of given examples.

In conclusion, the most promising changes brought by AM are the freedom to achieve complex geometric shape, material distribution, material composition, and function integration. When these changes come to the design process, they can be realized by structural optimization and function integration. However, both structural optimization and function integration happen in the downstream design flow. The need to explore the early design stage becomes urgent. The following issues are meant to be solved in the near future (Fig. 2):

1. Almost all the optimization design methods start with existing design which may jeopardize the potential of finding optimal design solutions because the original design is originally compromised.
2. The potential of AM in realizing function integration is seldom developed and most of the existing design are case-study-based. There is no theoretical framework to support function integration.
3. Although AM may help eliminate the need for assembly for some components, the constraints of any remaining assembly needs to set a new challenge for designing AM.

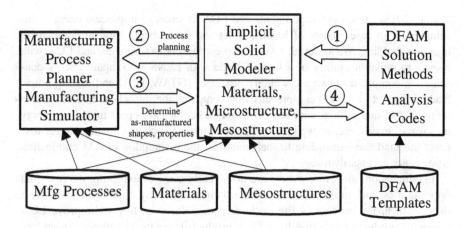

Fig. 2 DFMA system and overall methods (Rosen 2007). Adapted with permission

2.3 On-Going AM-Related Design Research on Sustainability

Additive manufacturing draws more and more attention for its great potential for sustainability because it has much improved materials efficiency, no-tooling manufacturing fashion, reduced life cycle impacts, and greater engineering functionality compared to subtractive manufacturing processes. As indicated by some researchers, 80 % of the environmental damage of a product is established after 20 % of the design activity is complete (Otto and Wood 1998). Therefore, identifying environmental impact factors in the early product development stage is critical.

AM-related design research on sustainability could be grouped into two categories on the basis of whether it is compared to conventional manufacturing processes. The first category is focused on the assumption that AM benefits the environment through a lightweight structure, less material consumption, and less energy loss. Huang et al. (2015) systematically estimated net changes in life cycle primary energy and greenhouse gas (GHG) emission for the adoption of metallic lightweight aircraft components fabricated by AM processes by the year 2050 to shed light on environmental impacts. In their study, it is indicated that cumulative energy savings could at most reach 1.2–2.8 billion GJ and GHG could have a reduction as high as 92.1–215.0 million metric tons. Some researchers (Mognol et al. 2006; Kellens et al. 2010; Baumers et al. 2011) also conducted similar quantitative studies on AM processes where only in-process energy consumptions are measured. Some researchers (Telenko and Conner Seepersad 2012) also include the energy consumption of raw material.

The second category concentrates on studying the difference between conventional manufacturing processes and AM in terms of detailed factors including material consumption, energy cost, hazardous material, and recyclability. Kreiger and

Pearce (2013) carried out experiments and LCA in terms of in-process energy consumption with case studies of blocks, spout and juicer being used to compare to injection molding. Wilson et al. (2014) also carried out experiments and LCA with respect to remanufacturing of a turbine blade with LENS to compare energy consumption compared to two arc welding processes (GTAW, PTA) and casting a new blade. Faludi et al. (2015) comprehensively compared the impact on environment in terms of AM and milling with respect to major ecological impact including energy use, waste, toxins, etc. as well as environment impact such as climate change, toxicity, and land use. According to their analysis, the assumption of AM eliminating waste is not necessarily true.

In conclusion, most researchers are at the early stage of exploring the potential opportunities and impact of AM on environment either by quantifying the effect of manufacturing process or life cycle analysis. However, how to improve LCA scores of products fabricated by AM in product development process is seldom studied. This study is therefore filling the spot of developing a design framework to secure product functionality with sustainability as a priority.

3 Sustainable Design Methodology for AM

From the brief review of AM-enabled design methods in the previous section, it is manifested that most existing design methods for AM are aiming at the improvement of products' functional performance. As to the sustainability of products fabricated by AM processes, most researchers are focusing on evaluating and minimizing the environmental impact of AM processes. However, it should be noted that those AM-enabled design methods may also play an important role for the sustainable product, because decisions made at the initial product design phase may also determine the environmental and economic impacts of future decisions (Harper and Thurston 2008). Thus, to reduce the product's environmental impact, it is necessary to link those AM-enabled design methods with the environmental impact evaluation model of AM processes. To achieve this goal, a general design methodology is proposed and described in this section. This general design methodology mainly focuses on the reduction of the environmental impact of the product's manufacturing process by taking advantage of AM technologies. It assumes the product's environmental impacts during other major life cycle stages are unchanged. This assumption is supported by a given design requirement which includes the restriction of the product's size, weight, or other parameters which may increase the product's environmental impact during other life cycle stages. In the following, the overall framework of the proposed methodology is first presented. Then four major design stages of this overall design framework are discussed, respectively, in each section.

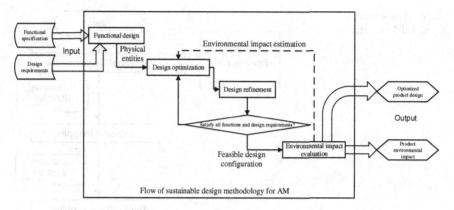

Fig. 3 General framework of sustainable design methodology for AM

3.1 General Design Flow

The general workflow of the proposed design methodology is shown in Fig. 3. The inputs of this design methodology are the product's functional specification and related requirements. In a functional specification, all major functions of a designed product should be declared, whereas in design requirements, the product's non-functional design constraints such as price, size, and weight are described.

The outputs of this proposed design method are an optimized design and its environmental impact. Generally, the whole design workflow can be divided into four stages—functional design, design optimization, design refinement, and environmental impact evaluation. In the functional design stage, physical entities are determined based on the product's functional specification. In the second stage, those AM-enabled design optimization methods can be used to minimize the product environmental impact. In this design stage, the feedback of environmental impact estimation needs to be considered. This feedback is characterized by the relationship between those design parameters of a product and its environmental impact. The product's environmental impact is considered to be one of the major design objectives during this design stage. After the design optimization stage, an initial feasible design solution can be generated. This initial design solution needs to be refined based on the manufacturability of the selected AM process, and the assembling ability of designed products also needs to be evaluated and considered. Some assembly features can be added. At the end of the design refinement stage, designers should check whether the refined design can satisfy all functions and design requirements from the input. If not, it should go back to the design optimization stage to modify the design parameters. Otherwise, it can go to the environmental impact evaluation stage where an environmental impact evaluation model can be applied to calculate the environmental impact of product's manufacturing process.

Fig. 4 Design flow of
functional design stages

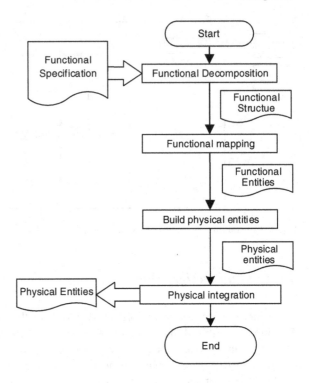

3.2 Functional Design

The general design steps of the functional design stage are shown in Fig. 4. The
major input of this design stage is a functional specification. A functional specification usually defines the overall functions of a product and some requested input
and output properties. It usually does not define the internal working process of
a product. Generally, designers can summarize the interaction between designed
products and its external agent, e.g., users, material, and energy, based on the input
functional specification. A generic black box model can be used to represent the
input functional specification which is shown in Fig. 5.

For those products that only play a single or several basic functions, it is not
difficult to find directly their corresponding physical features to fulfill their functional requirements. However, if designers cannot find directly feasible physical
features to meet directly the overall functions of a product, a functional decomposition process is needed. In this chapter, functional decomposition refers to a
process which resolves the overall functions of a designed product into a set of

Fig. 5 Generic black box
model for product functional
specification

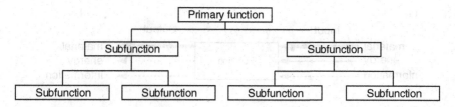

Fig. 6 Generic format of functional tree

interlinked subfunctions in such a way that the original functions can be fulfilled by implementing all subfunctions. These subfunctions and their interlinked relationship are defined as a functional structure which is the output of a functional decomposition process.

An elementary approach for functional decomposition is to decompose hierarchically the input functions of a designed product into a tree structure, usually known as a functional tree (shown in Fig. 6). The functional tree is easy and fast to build. However, it only contains hierarchical relationships between subfunctions, and fails to describe the interaction between different functions. To describe the interaction between different subfunctions and their relations to the overall functions of a product, a black box functional model (shown in Fig. 7) can be used in the functional structure.

In this functional structure, a basic function is described with a simple black box whose input and output are clearly defined, as shown in Fig. 7a. Different basic functions are interlinked according to the energy, material, and information flow. The generic format of this type of functional structure is shown in Fig. 7b. Compared to the functional tree, the functional structure based on a black box model contains more information and also needs more time to construct. This type of functional structure can be an efficient tool for designing a product with complex functional interaction. Generally, according to the functional complexity of a designed product, designers can select an appropriate functional structure. For the detailed steps of constructing different types of functional structure, the reader can refer to (Otto and Wood 2001).

After functional decomposition, a functional structure can be obtained. A functional mapping is needed to map the obtained functional structure into physical features and its relations. At the current stage, the referred physical feature is not assigned with detailed geometry and material information. It is an abstract entity only with the information of its physical behavior which can be used to implement its corresponding function. Thus, these entities obtained from functional mapping are called functional entities. These functional entities are usually searched based on designers' experience or knowledge in the current design step. However, because AM-enabled design features have not been widely used, designers usually lack awareness of these features. This can be a barrier to taking those AM-enabled features to reduce environmental impact. To overcome this barrier, an

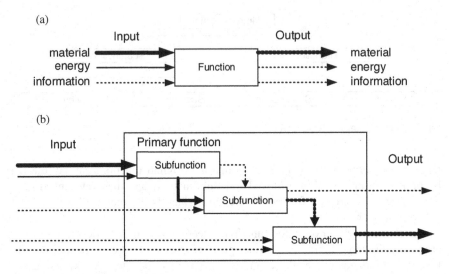

Fig. 7 Generic format of functional structure based on a *black box*. **a** *Black box* of basic function. **b** Generic format of functional structure

AM-enabled design feature database or knowledge base is needed. For example, an AM-enabled feature database has been built by Maidin et al. (2012) to inspire designers.

It should also be noted that the relationships between basic functions in a functional structure and their corresponding functional entities are not always in one to one correspondence. Indeed, one basic function can always find one corresponding functional entity. However, one functional entity is not necessary to serve only one basic function. In some cases, a functional entity can serve several basic functions simultaneously. The process of mapping several basic functions into one functional entity is known as functional integration. An engine of a car is an example of functional integration. In the functional design of a car, an engine can be regarded as a functional entity. It serves two different functions. First, it can transfer the fuel energy into kinetic energy. Second, it also generates heat for the heating system. Thus, it can also play a role as a heater. Obviously, functional integration can reduce the number of functional entities needed, which may lead to the reduction of the overall parts' count. However, it may also cause some issues. One of the most serious issues that functional integration may bring is the functional coupling. These basic functions are coupled when they are all served by the same functional entity. The design parameters of this functional entity may affect the function performance for different function simultaneously. Thus, difficulties may arise in the following design steps in deciding the detailed design parameters of this functional entity. Thus, functional integration is not suggested by some existing design methodologies, such as axiomatic design theory (Suh 1998).

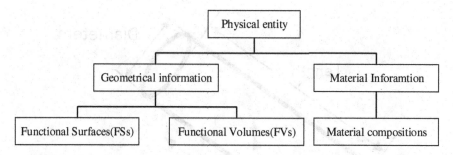

Fig. 8 Graphic view of data structure for physical entity

Fig. 9 FS and FV of airfoil

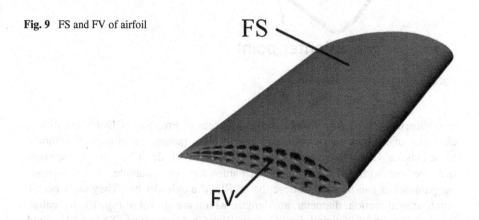

The next step of functional design is to construct concrete entities to realize the physical behavior described by obtained functional entities. This concrete entity should contain two types of information, geometrical information and material information. In this chapter, a concrete entity built in the current step is referred to as a physical entity. The graphic view of a physical entity's data structure is shown in Fig. 8. To represent the geometrical information of a physical entity, concepts of Functional Surfaces (FSs) and Functional Volumes (FVs) are used. In this chapter, an FV is defined as a geometrical volume of functional entity, whereas an FS is a key surface of a functional entity for its physical behavior. For example, Fig. 9 shows the physical entity of an airfoil. The outer surface of this airfoil is the key surface which plays an air dynamic role. The whole structure of this airfoil is the FV. In this FV, a lattice structure is used to reduce its weight.

In the current design stage, because of the incomplete information grasped by designers, it is impossible to make a final decision on the exact shape of FSs and FVs. Thus, the defined FSs and FVs at the current stage are changeable and deformable surfaces or volumes. A parametric modeling method can be used to describe those deformable surfaces or volumes. In this chapter, a parameter vector θ is used to control the shape of FSs or FVs. The set of all allowable value for the parameter is denoted $\Theta \subseteq \mathcal{R}^k$ where k is the dimension of a parameter vector.

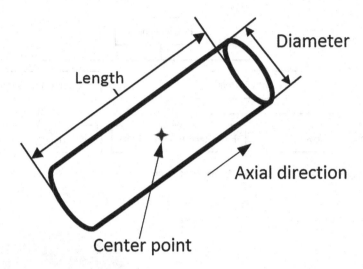

Fig. 10 FV in cylindrical shape

This dimension is also known as Design Degrees of Freedom (DDoF). For different types of geometry element, there are different parametric modeling methods. Most existing CAD software provides the capability to describe a simple geometrical element with several independent parameters. For example, four independent parameters can be used to describe an FV of a cylinder bar. They are a center point, axial direction, diameter, and length, which are shown in Fig. 10. To realize its corresponding functional requirements, some parameters of FSs and FVs need to be fixed. For instance, if the FV shown in Fig. 10 is designed to fit a hole with a certain diameter and axial direction, both diameter and axial direction of this FV should be fixed with given values. Thus, the DDoF of this FV is two.

Because of the constraints of traditional manufacturing, designers at the current stage traditionally tend to assume FVs and FSs in a simple geometry with the small number of DDoF. These assumptions can greatly reduce the complexity in following design processes. Moreover, the product can be generated with regular geometry, which is easy manufacturing. However, whether these FSs and FVs are optimized with respect to functional performance or environmental impact is hard to decide. For example, Fig. 11 shows a design case of a physical entity with one FV and two FSs to sustain a normal pressure P on surface C with fixed end at surface A. To realize this physical behavior, most experienced designers may select the "I" shape beam as an FV for this physical entity to sustain the bending moment. However, the result of topology optimization shows the irregular truss-like shape structure may achieve the same stiffness with less material than the regular "I" shape. Thus, to take advantage of design freedom provided by AM technologies, the parametric modeling methods, which can deal with complex geometrical shapes, are needed to describe the FSs and FVs of physical entities which are to be fabricated by AM processes. For example, a complex FS

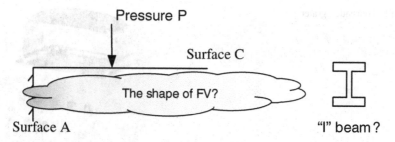

Fig. 11 Simple FV versus complex Fv

can be represented by a NURBS (Non-Uniform Rational B-Splines) surface. The positions of control points can be regarded as parameters to describe this FS. As to FV, a voxel-based geometrical modeling method can be used to describe the complex geometry of FV. The scalar value at each voxel point is the parameter to control the shape of FV. If this value is larger than zero, this voxel point is considered in the FV, or else this voxel point is out of FV. Indeed, using an FS or FV with more DDoFs may increase the complexity in the following design process. However, those AM-enabled design methods discussed in Sect. 2 are able to deal with a large number of DDoFs to generate an optimized result with respect to both functional performance and environmental impact.

Besides geometrical shape, an appropriate material is also needed to be decided for each physical entity at the current step. The classical material selection method based on a material chart (Ashby and Cebon 1993) can be used here to help designers select the material which can achieve the required physical phenomenon with the minimum environmental impact.

At the end of the functional design stage, to reduce further the product's part counts, some physical entities can be merged into one physical entity. This process is known as physical integration. Based on the general capability of AM processes, some simple rules for physical integration are provided in this chapter. The two physical entities satisfying all these rules can be considered as candidates for physical integration.

Rule 1: There is no relative movement between two physical entities.

Rule 2: Material of two physical entities is compatible with respect to a certain manufacturing process.

It should be noted that the two rules mentioned above are only necessary conditions that integrated physical entities should have. More detailed manufacturing and assembly information is also needed for designers to make the final decision. For those integrated entities, the FSs which play roles as connection surfaces between them can be removed, and the connected FVs can be merged together. One example of physical integration is a four-sided grater which is shown in Fig. 12. To design this product, each side of grater itself is originally the physical entity which is used to grate the vegetable into different shapes. In this step, designers can combine these four independent physical entities into one part. It is clear that their

Fig. 12 Four-sided grater

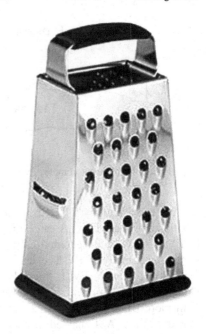

original FSs for grating are still kept and independent. However, the overall part count has been decreased. As with the functional integration process mentioned during the second step of the current design stage, the physical integration process can also reduce the overall part count of a designed product. However, the functional entities and their related FSs and FVs still remain independently inside the integrated physical entity. Thus, functions implemented by this physical entity are still decoupled. The physical entities built at the end of the functional design stage are regarded as the input in the following multiscale design optimization stage.

3.3 Design Optimization

In the second stage, a design optimization process can be applied to the physical entities obtained to minimize product environmental impact while improving its functional performance. The design parameters of FSs and FVs are regarded as the design variables of this optimization process. Moreover, the pre-feedback of the environmental impact model is considered with those multiscale AM-enabled design optimization methods described in Sect. 2. Its general work flow is shown in Fig. 13.

This workflow can be divided into four steps. At the beginning, the objective of the optimization process is determined based on the pre-feedback from the environmental impact model. Then functional requirements and manufacturing constraints are converted to the constraints on the design parameters. After that, according to the major function played by a designed physical entity, an

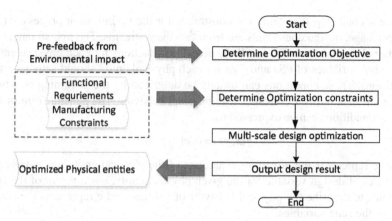

Fig. 13 General work flow of optimization process

AM-enabled design optimization method can be select to optimize this physical entity on different design scales. At the end, the design result is output to the next design stage for design refinement. In the following paragraphs, the detailed design steps of this design stage are discussed in detail.

First of all, to determine the objective of the following optimization process, the pre-feedback of the environmental impact model on the selected AM manufacturing process needs to be considered. The detailed discussion on the environmental impact model of the AM manufacturing process is given in Sect. 3.5. In the optimization process discussed in this section, the environmental impact evaluation model is considered as a function which can expressed as

$$I_e = f\left(p_{\text{design}}, p_{\text{machine}}, p_{\text{material}}, p_{\text{operation}}\right) \tag{1}$$

where I_e is a vector of environmental impact indexes such as midpoint indicators of ReCiPe, $p_{\text{design}}, p_{\text{machine}}, p_{\text{material}}, p_{\text{operation}}$ are the vectors of design-dependent, machine-dependent, material-dependent, and operation-dependent parameters for the environmental impact evaluation model, respectively. The function shown in (1) only depends on the type of AM process selected. For different types of AM processes, the form of function might be different.

During the optimization of each physical entity, the type of manufacturing process needs to be predetermined based on the selected materials and the shape of FSs and FVs. Once the type of AM process is determined, the form of function f shown in (1) can be obtained. Based on this function, the first round of the minimization process can be used to find the design-dependent parameters p_{design}^*, which can achieve the minimum environmental impact. During the optimization process, other independent variables in function f are unchanged. It should be noted that some elements in the vector in the p_{design}^* might equal zero or infinity. This value is regarded as the pre-feedback of the environmental impact model on the design process. This pre-feedback can be set as the design objective during the following optimization process.

The second step is to build the constraints for the optimization process. At the current stage, design constraints are from two main streams. The first stream is the functional requirements. To fulfill the described functions for each physical entity, the design variables of FSs and FVs for each physical entity should satisfy certain conditions. These conditions can usually be described by the governing equations of the physical entity's corresponding physical behavior. The general form of this type of conditions can be expressed as

$$g(x, y) = 0, c(y) > 0 \qquad (2)$$

where g represents the governing equations of the physical entity's behavior, x is the vector of design variable for the given physical entity, y is a vector of the state variables to describe the physical behavior of entities, and c represents the conditions of the state variables.

Besides those functional related constraints, the capability of the selected AM process is another main stream for design constraints. Although the AM process can fabricate a part with an extremely complex shape, it still has certain limitations. For example, support structures are needed for certain types of AM processes and some of these support structures are difficult to remove because of inaccessibility. Thus, during the optimization process of physical entities, the manufacturing constraints also need to be considered.

Based on the obtained design objective and constraints, the design can be described so as to find design variables for physical entities which can achieve the optimized design-dependent parameters p_{design}^* while satisfying all the functional requirements. Thus the optimization problem can be expressed as

$$
\begin{aligned}
&\text{Min. } |p_{\text{design}}(x) - p_{\text{design}}^*| \\
&\text{S.T. } g_i(x, y) = 0, \; c_i(y) > 0, \; i = 1, 2, 3 \ldots l \\
&\qquad m_i(x) > 0, \; j = 1, 2, 3, \cdots k
\end{aligned} \qquad (3)
$$

where $|n|$ denotes a norm of vector n, x represents a vector of design variable for physical entity, $p_{\text{design}}(x)$ is a function which can map design variables of physical entities into the design-dependent parameters for environmental impact model, $g_i(x, y) = 0$, $c_i(y) > 0$ is the constraint from the ith function of the physical entity, and $m_i(x) > 0$ is the constraint from manufacturability of a selected AM process.

To solve the optimization problem stated in (3), various AM-enabled design optimization methods described in Sect. 2 can be used. However, most existing design optimization methods focus on improving the functional performance of the designed physical entity. Thus, sometimes it is difficult to apply directly those optimizations to solve the optimization problem defined in Eq. 3. To deal with this problem, designers can convert the objective function in (3) into the constraints of structural optimization problems. Consider a sequence of design-dependent parameters p_{design} which can be denoted as $L = (p_1, p_2, \ldots p_n, \ldots, p_m, \ldots)$. This sequence of design-dependent parameters satisfies the following condition: for any n, m, if $n < m$, then $|p_n - p_{\text{design}}^*| > |p_m - p_{\text{design}}^*|$. Based on this sequence of design-dependent parameters, the optimal p_{design} can be searched by the algorithm presented in Fig. 14.

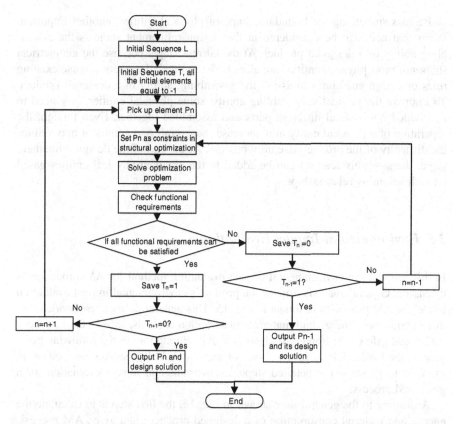

Fig. 14 Algorithm to find the optimized design solution for minimum environmental impact based on existing structural optimization methods

3.4 Design Refinement

After the design optimization stage, a design refining process is needed to modify further some detail features of optimized product design because of the coarse boundary or irregular boundary obtained from the second design stage. Especially for those physical entities designed with the relative density based topology optimization method, the result of the optimization process is a relative density distribution in the FVs. Thus, at the design refinement stage, designers need to choose a method to deal with gray regions where the relative density is between 0 and 1. The simplest way is to set a threshold of relative density. In the region where the relative density is lower than this threshold, the material is removed. This method is simple but may cause some unforeseen problems. For example, if the threshold is too small, the optimized FV of physical entity might be divided into several separated portions. Otherwise, the big threshold leads to too much material left at the end, which cannot achieve the optimized design. Thus, designers should be careful to check the threshold at the design refinement stage.

Besides smoothing the boundary shape of physical entities, another important factor that needs to be considered at the design refinement stage is the assembling ability of a designed product. At the current stage, because the geometrical shapes of most physical entities are already determined, designers can use existing rules or design guidelines to assess the assembling ability of a designed product. To improve the product's assembling ability, some physical entities may need to be divided into several different parts and assembled together. Even though, the separation of a physical entity may increase the overall part counts, it may reduce the difficulty of the product assembly process. At the end of the design refinement stage, the assembly features can be added to the obtained physical entities based on their assembly relationships.

3.5 Environmental Impact Evaluation

In this section, the environmental impact evaluation method for AM processes is discussed. General analysis flow of the product's environmental impact evaluation model for AM processes is shown in Fig. 15. This general flow can be divided into three main steps: energy and material consumption analysis, Life Cycle Inventory (LCI), and Life Cycle Impact Analysis (LCIA) compilation. In the following paragraphs, the binder jetting process, one of the major AM processes, is used as an example to illustrate the detailed steps of environmental impact evaluation for a given AM process.

According to the general flow shown in Fig. 15, the first step is to calculate the energy and material consumption of a designed product via a given AM process. To achieve this purpose, the manufacturing process of the selected AM technique is first analyzed. Generally, the whole manufacturing process of a binder jetting technique can be divided into four steps: printing, curing, depowdering, and sintering. The core step of the binder jetting process which differentiates it from other AM technologies is the printing process. It is difficult to evaluate directly energy and material consumption of the printing process, because this process consists

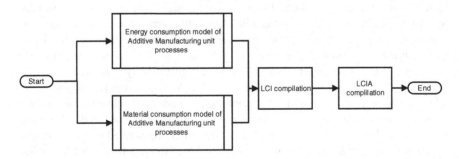

Fig. 15 Workflow of environmental impact evaluation

Fig. 16 Working principle of printing process for the binder jetting technique

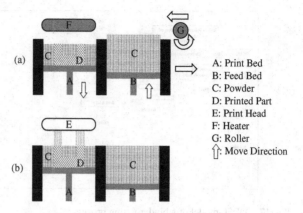

A: Print Bed
B: Feed Bed
C: Powder
D: Printed Part
E: Print Head
F: Heater
G: Roller
⇧: Move Direction

of three subprocesses: spreading, printing, and heating. As showed in Fig. 16a, a printing process starts by lowering the print bed by one layer thickness and lifting the feed bed by one layer thickness. Then the roller spreads one layer of powder materials from the feed bed to the print bed. This process is known as spreading. Then the print head E deposits a pattern onto the powder with binder material, thus forming a printed layer. This subprocess is known as printing a binder. After one layer is finished, the step motor system moves the print bed under an electrical infrared heater to dry the binder. This subprocess is known as heating. After one layer of printing, the machine automatically repeats this process until the part is completed.

Besides these subprocesses, the printing preparation process also needs to be considered in the binder jetting process. To summarize the manufacturing processes mentioned above, the IDEF0 model of the binder jetting process is established and shown in Fig. 17. Based on the IDEF0 model of the binder jetting process, an LCA process model is built on UMBERTO NXT LCA software which is shown in Fig. 18. The parameters related to subprocesses of LCA process model shown in Fig. 18 are predefined. These parameters can be divided into four types: machine-dependent parameters, operator-dependent parameters, material-dependent parameters, and design-dependent parameters. These four types of parameters are summarized in Tables 1, 2, 3, and 4 respectively.

For each of the entities modeled in UMBERTO NXT LCA software, a mathematical expression for each activity is first developed. The detailed discussion of this mathematical modeling can be found in (Meteyer et al. 2014). Here, only a summary of those mathematical models is listed.

Energy consumption models:

Infra-red heater

The infra-red heater power is set as a percentage of its maximum power by the operator on the machine and is running during the entire process. The heating

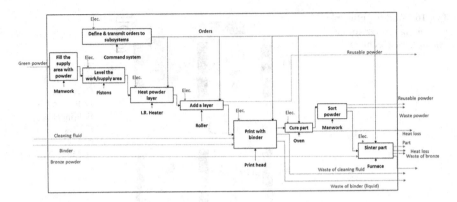

Fig. 17 IDEF0 model of a binder jetting process

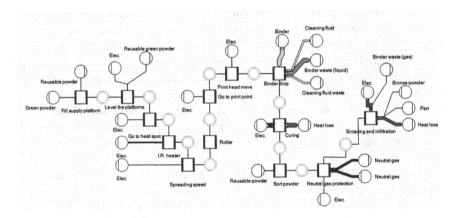

Fig. 18 The LCA unit process model of binder jetting unit process

time of the powder is defined by the time the platform stays under the heater. The
energy used by the heater is defined by

$$E_{heater} = \frac{\left(P_{max} \times \%heater \times t_{process} \right)}{Eff_{heater}} \qquad (4)$$

with

$$t_{printing} = \frac{H_{part} \times t_{layer}}{H_{layer}} \qquad (5)$$

The maximum electrical power used by the heater is measured and the maximum power of the heater is found in the machine documentation.

Table 1 Machine-dependent parameters

Machine-dependent parameters		
Parameter	Symbol	Unit
Volume of supply platform	V_{supply}	mm^3
Screw diameter	D_{Screw}	mm
Mass of platforms	M_{platform}	kg
Platforms' section	S_{platform}	mm^2
Screw's pitch	p	mm
Efficiency of transmission for supply and part build platforms	$\eta_{\text{transmsupply/partbuild}}$	%
Efficiency of motors for supply and part build platforms	$\eta_{\text{motsupply/partbuild}}$	%
Friction coefficient in screw-nut systems	$f_{\text{screw - nut}}$	Unit
Mass of chariot	M_{chariot}	kg
Friction coefficient in chariot's guiding	$\mu_{\text{slideways}}$	Unit
Distance print spot-heat spot	$L_{\text{print−heat}}$	mm
Efficiency of pulley-belt for roller	$\eta_{\text{pulley - belt}}$	%
Efficiency of motor for chariot	$\eta_{\text{motroller}}$	%
Maximum power of infrared heater	$P_{\text{heatermax}}$	W
Furnace volume	V_{furnace}	mm^3
External pressure	p_{ext}	Pa
Pump's mechanical efficiency	η_{pump}	%
Distance spreading	L_{spread}	mm
Distance end of spread spot-print spot	$L_{\text{spread - print}}$	mm
Print-head stroke	L_{phstroke}	mm
Mass of print head	$M_{\text{print - head}}$	kg
Friction coefficient in print-head guiding	$\mu_{\text{print - head}}$	Unit
Efficiency of rack and pinion for print-head	$\eta_{\text{r\& p}}$	%
Power of uncapping	P_{uncap}	W
Specific heat capacity of the apparatus for curing	$Cp_{\text{apparatus}}$	J/kg/K
Specific heat capacity of the recipient for sintering	$Cp_{\text{recipient}}$	J/kg/K
Specific heat capacity of the support powder	$Cp_{\text{supportpowder}}$	J/kg/K
Specific heat capacity of the infiltrant	$Cp_{\text{infiltrant}}$	J/kg/K
Surface of oven	S_{oven}	mm^2
Convection coefficient inside the oven	h_{intoven}	W/m2/K
Convection coefficient outside the oven	h_{extoven}	W/m2/K
Convection coefficient inside the furnace	$h_{\text{intfurnace}}$	W/m2/K
Convection coefficient outside the furnace	$h_{\text{extfurnace}}$	W/m^2/K
Thermal resistance of oven wall	R_{oven}	m^2 K/W
Thermal resistance of furnace wall	R_{furnace}	m^2 K/W
External temperature	T_{ext}	K
Mass of recipient for sintering	$M_{\text{recipient}}$	Kg

Table 2 Operator-dependent parameters

Operator-dependent parameters		
Parameter	Symbol	Unit
Percentage of filling supply platform	$\%_{filling}$	%
Mass of reused powder	M_{reused}	kg
Layer thickness	δ	mm
Feed ratio	R_{feed}	Unit
Percentage of heater's maximum power	$\%_{heater}$	%
Vacuum pressure desired	p_{fin}	Pa
Argon flow-rate	D_{vAr}	mm^3/s
Mean time between two consecutive layers	t_{layer}	s
Number of overlaps	$N_{overlaps}$	Unit
Saturation ratio	R_{sat}	
Mass of binder waste per layer	$M_{binder/layer}$	kg
Mass of cleaning fluid waste per layer	$M_{clean/layer}$	kg
Mean time to print a layer	$t_{printlayer}$	s
Mean temperature during curing	$T_{meancuring}$	K
Total duration for curing	$t_{totalcuring}$	s
Duration of maintain phase for curing	$t_{maintaincuring}$	s
Mean temperature during sintering	$T_{meansintering}$	K
Total duration for sintering	$t_{totalsintering}$	s
Duration of maintain phase for sintering	$t_{maintainsintering}$	s
Infiltrant ratio	$R_{infiltrant}$	Unit

Table 3 Material-dependent parameters

Material-dependent parameters		
Parameter	Symbol	Unit
Density of powder	ρ_{powder}	g/mm^3
Packing ratio	$\%_{pack}$	%
Proportion of reusable powder	$\%_{reusable}$	%
Density of binder	ρ_{binder}	g/mm^3
Specific heat capacity of the powder	Cp_{powder}	J/kg/K

Table 4 Design-dependent parameters

Design-dependent parameters		
Parameter	Symbol	Unit
Height of part	h_{part}	mm
Volume of part	V_{part}	mm^3

Curing

The curing oven is modeled as a hermetically closed oven the walls of which are made of one material of surface thermal resistance R. Estimation of this resistance is discussed later. The curing profile consists of a linear increasing of the temperature and a maintained period for the final temperature. The energy needed is split into the energy for heating the powder and the apparatus and the energy needed to maintain the temperature in the oven.

$$E_{heating} = \left(M_{powder} \times Cp_{powder} + M_{support} \times Cp_{support}\right) \times \Delta T \qquad (6)$$

$$E_{maintain} = \int_0^{t_{fin}} \frac{S_{oven} \times (T(t) - T_{ext})}{R + \frac{1}{h_{int}} + \frac{1}{h_{ext}}} \times dt \qquad (7)$$

$$E_{maintain} = \frac{S_{oven} \times (T_{mean} \times t_{increase} - T_{maintain} \times t_{maintain})}{R_{oven} + \frac{1}{h_{int}} + \frac{1}{h_{ext}}} \qquad (8)$$

Sintering

The sintering profile consists of several linear temperature increases followed by a maintained temperature. By analogy, the sintering energy is defined as

$$E_{heating} = (M_{powder} \times Cp_{powder} + M_{support} \times Cp_{support} + M_{supportpowder}$$
$$\times Cp_{supportpowder}) \times \Delta T \qquad (9)$$

$$E_{maintain} = \frac{S_{oven} \times \left(T_{mean} \times t_{increase} - \sum (T_{maintain} \times t_{maintain})\right)}{R_{furnace} + \frac{1}{h_{int}} + \frac{1}{h_{ext}}} \qquad (10)$$

Idle state energy

The main source of energy consumption of the printing has been found to be the idle state energy consumption, consisting of computer consumption (60 W), lighting consumption (12 W), and other sources such as controllers (50 W). These elements are running during the entire printing process which explains their importance in the general energy consumption of the machine.

Others

Experiments have shown that all the other component's energy consumption represents less than 1 % of the total energy consumption. These components are, however, included in the model in view of further studies, but are not described in this chapter.

Material consumption models

Print with binder

Because of its viscosity characteristics, binder has to be washed out of the system frequently. These cleanings are made every two layers with the M-Lab machine.

The binder and cleaner consumption are therefore linear with the number of layers:

$$M_{\text{binder}} = \frac{\rho_{\text{binder}} \times Vb_{\text{layer}} \times H_{\text{part}}}{H_{\text{layer}}} \tag{11}$$

$$M_{\text{cleaner}} = \frac{\rho_{\text{cleaner}} \times Vc_{\text{layer}} \times H_{\text{part}}}{H_{\text{layer}}} \tag{12}$$

Powder

In this study, all the powders used in the process but not printed are considered reusable because unused powders are combined with new powders for new rounds of printing. No significant mechanical property change has been observed when old and new powders are used together. The overall amount of reusable powder is therefore the mass of powder used to fill the supply system minus the mass of the part.

Once the binder jetting AM process is modeled in the UMBERTO NXT LCA system, the LCI data can be calculated with defined reference unit and reference flow. Then some inventory databases such as Ecoinvent v3 (Weidema et al. 2011) can be used for secondary material and energy consumption evaluation such as powder manufacturing and binder manufacturing. Based on the LCI data, some indicators such as ReCiPe midpoint (Goedkoop et al. 2008) are chosen to assess the environmental impact generated through the manufacturing processes. The environmental impact result is the output at the end of the proposed design method.

4 Case Study

In this section, a case study is provided to illustrate further the proposed design methodology. In this case study, a triple clamp of a motor cycle is used. The original design of this product is shown in Fig. 19.

To redesign this product, the functional specification and design requirements are first summarized. The primary function of a designed triple clamp is given below:

Function 1: To connect steering handle and front fork with motorcycle frame (shown Fig. 20). This connection can transfer torque from the steering handle to the front fork which allows the front fork to pivot from side to side.

The design requirements of a designed triple clamp are also listed below:

Design Requirement 1: The solid connection should be achieved by a designed product, which means a triple clamp does not fail or break during its working state.

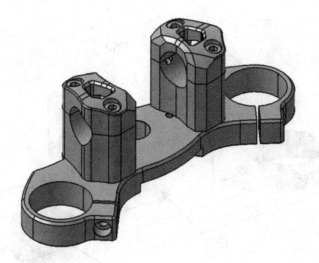

Fig. 19 Triple clamp of a motor cycle

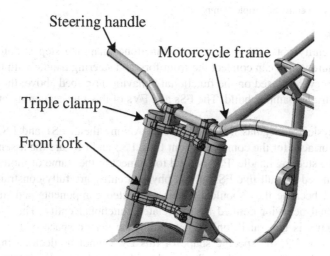

Fig. 20 Primary function of a triple clamp

Design Requirement 2: The connection should be stiff and rigid enough, which means the maximum deflection of a design triple clamp should smaller than a given value.

Based on the input functional specification and design requirements, the proposed design methodology is applied to redesign this triple clamp during the functional design stage, because the primary function of a designed product is easy to achieve. Thus, the functional entity can be directly obtained from the primary function of a triple clamp without a functional decomposition process. The

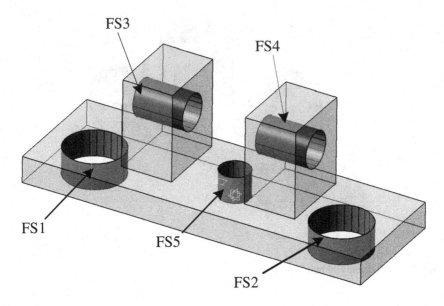

Fig. 21 Physical entity of a triple clamp

functional entity obtained at the end of functional mapping step is defined as a solid structure which can connect the front fork and steering handle with the frame of the motorcycle. Based on the functional behavior described above, the physical entity of a triple clamp is build. The FSs and FVs of this physical entity are shown in Fig. 21.

This physical entity has five FSs in total. Among them, FS1 and FS2 are the assembly surfaces for the connected front fork. FS3 and FS4 are the assembly surfaces for the steering handle. FS5 is used to connect to the frame of motorcycle. It should be noted that all five FSs of this physical entity are fully constrained with zero DDoF, because they should fit their connected components and implement the functional behavior defined by the related functional entity. The FV of this physical entity is generated, only representing the design space of FV. As mentioned in Sect. 3.2, the specific shape of this FV cannot be decided in the current step. In this case study, the redesigned product is planned to be fabricated by the AM process. In order to take the unique capability of the AM process which can fabricate parts with complex geometry, the FV of this physical entity is represented by the voxel-based parametric modeling method. The DDoF of this FV is equal to the number of voxel points needed to represent the design space of an FV. For this case study, the size of the voxel point is chosen as 3 mm according to the dimension of an FV.

Besides the geometrical shape of a designed physical entity, the material of this physical entity also needs to be determined. In this design case, stainless steel is used for the original design. This material can also be used for a redesigned product. However, the mechanical properties of printed stainless steel may be slightly

Table 5 Mechanical properties of printed stainless steel fabricated by a binder jetting process

Material	Elastic modulus	Ultimate strength	Density	Yield strength
Stainless Steel 316 infiltrated with bronze	148 GPa	407 MPa	7.86 g/cm^3	234 MPa

different from the properties of the stainless steel fabricated by traditional manufacturing processes such as milling. For this case study, the redesigned product is supposed to be fabricated by a binder jetting process. Some basic material properties of printed stainless steel 316 are listed in Table 5 (E. Inc 2014). These properties are used in the design optimization stage, which is discussed in the following paragraphs.

For this case study, there is only one physical entity which serves for one functional entity. Thus, the physical integration is no longer needed. After the functional design stage, the physical entity shown in Fig. 22 can be output for the design optimization stage. In the design optimization stage, the feedback of environmental impact evaluation needs to be calculated first. According to the energy and material consumption model described in Sect. 3.5 for the binder jetting process, there are two design-dependent parameters. They are part volume V_{part} and height of a part h_{part}. As to the height of a part, it is almost constrained by the position relationship of FSs. For this design case, the DDoF of FSs is equal to zero. Thus, in this design process, the volume of the designed product is regarded as the design target. By analyzing the environmental impact model, the minimal environmental impact can be achieved when the design-dependent parameter V_{part} equals zero. Thus, the p_{design}^* can be determined as below:

$$p_{design}^* = (V_{part}) = (0) \tag{13}$$

After determination of the design objective, the topology optimization method is used to update the relative density for each voxel point in FV to obtain the final design result. In the topology optimization process, two different load cases (shown in Fig. 22) are considered based on the existing literature (Kumar and Choudhary 2015) for the triple clamp design. Load case 1 is the steering torque applied on a triple clamp. Load case 2 is the vertical impact force applied on a triple clamp. For both load cases the FS1, FS2, and FS5 are constrained with six degrees of freedom.

Besides the boundary conditions discussed above, in the topology optimization process the thin layer of material should be kept around the mentioned FSs for assembly purposes. Thus, thin layers of material around FSs are denoted as non-design space for topology optimization. The thickness of these thin layer material is 2 mm for this design case. The design space and non-design space of this case study are shown in Fig. 23.

Fig. 22 Load condition of a
designed triple clamp

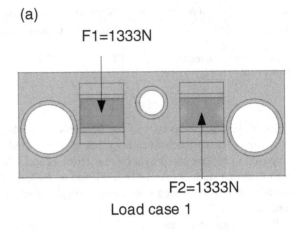

(a)

Fig. 23 Design space of
topology optimization

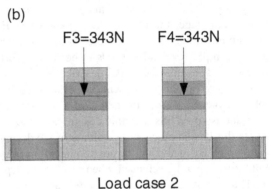

(b)

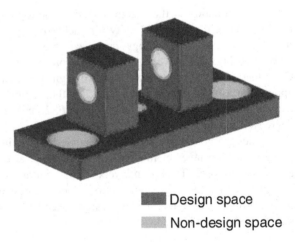

Fig. 24 Result of topology optimization

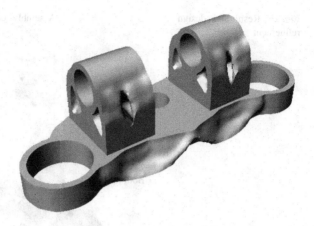

The objective of topology optimization is the overall stiffness of a designed product rather than product volume. Thus, a sequence of product volume fraction is generated first. In this volume fraction sequence, the volume fraction ranges from 10 % to 100 % at 10 % intervals. The elements in this volume fraction sequence are regarded as the constraints during the topology optimization. Besides volume fraction constraints, the yield strength of printed stainless steel is also regarded as a design constraints during the optimization process because of the Design Requirement 1 mentioned above. The topology optimization problem is solved by OptiStruct solver (Engineering 2009). Based on the algorithm described, the minimum volume fraction of a designed triple clamp which can satisfy all design requirements is found to be 40 %. The optimized result under this volume fraction constraint is shown in Fig. 24.

After the design optimization step, design refinement is needed to smooth the boundary of FV obtained from the last design stage. Moreover, the assembly ability also needs to be evaluated. It is difficult to assemble the steering handle to the triple clamp with the current design shown in Fig. 24. To ease the assembly process, the original design is broken down into two sections with three parts. Moreover, some assembly features are also added in the optimized design. The result of the design refinement stage is shown in Fig. 25.

At the end of the design process, the environmental impact evaluation model is applied to the obtained design. To compare this value with its original design, the environmental impact factor of the original design is also calculated based on the environmental impact evaluation model of the milling process from UMBERTO NXT LCA software. For comparison, the following information is kept the same:

1. The scope of the LCA study is to compare the environmental impact of different design solutions generated based on functional design methods. It is also in the scope to compare the environmental impact of different manufacturing methods specifically between the conventional milling process with the binder jetting AM process. The function unit is a single design product. The reference flow is the required service life expectancy. Different design solutions all have the same expected service life expectancy.

Fig. 25 Result after design
refinement

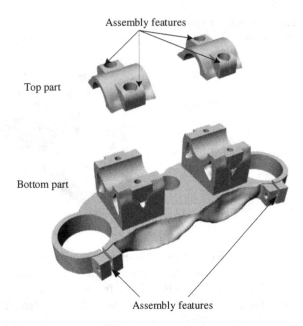

2. Assumptions are made in the LCA study. It is assumed that the conventional
   CNC milling process is a representative manufacturing method to fabricate the
   design solutions. All features of the design can be successfully machined without failure. It is also assumed that the binder jetting process is able to manufacture the design solution successfully without failure.
3. The boundaries of this LCA study are set only to consider the electricity needed
   for the chosen function unit and reference flow; the environmental impact produced from manufacturing electricity generation equipment is cut out. The
   environmental impact produced from binder jetting manufacturing and conventional CNC machines are also cut out.

The comparison between the original product and the optimized product is made.
In this case study, the ReCiPe midpoint indicator is used to analyze quantitatively
the environmental impact of the designed products. Some of the major ReCiPe
midpoint indicators obtained from the environmental evaluation model are shown
in Table 6. It is clear that the redesigned product has less environmental impact
than that of its original design fabricated by traditional manufacturing processing.
Moreover, the overall parts count of a designed product is also reduced. A comparison of the parts count between the original design and the optimized design
is shown in Table 7. It is obvious that the proposed design methodology can help
designers to reduce the overall parts count, which can also make a contribution to
less environmental impact and less cost.

The proposed case study discussed in this section shows the unique capability of the proposed sustainable design methodology. It is clear that by redesigning

Table 6 ReCiPe midpoint of the designed product

	Binder Jetting	Milling
Agricultural land occupation	0.77 m^2	1.77 m^2
Climate change/CO$_2$	3.13 kg	30.72 kg
Fossil depletion	1.68 kg	6.24 kg
Freshwater ecotoxicity/FETP100/1,4-DCB-Eq	0.01 kg	1.44 kg
Human toxicity/1,4-DCB-Eq	0.06 kg	3.06 kg
Ionizing radiation, IRP_I/U235-Eq	0.16 kg	1.96 kg
Marine ecotoxicity/1,4-DCB-Eq	8.81E−3 kg	1.33 kg
Marine eutrophication, MEP/N-Eq	3.88E−3 kg	0.06 kg

Table 7 Comparison between original design and optimized design

	Parts count (including assembly bolts)	Volume/cm^3
Original design	13	4.14e10^2
Optimized design	7	3.44e10^2

the existing product with the proposed design methodology the environmental impact of product's manufacturing process can be significantly reduced. The reduction is mainly because of the AM-enabled design method, topology optimization, used in the proposed design methodology. Moreover, the overall part count is also decrease from 13 to 7. Generally, by taking the advantage of the AM process, the proposed design methodology in this chapter enables designers to minimize the product's environmental impact of its manufacturing process as well as to reduce further its overall parts count through functional integration and parts consolidation.

5 Summary

In this chapter, a sustainable design methodology for the products fabricated by AM process is discussed. First, the current state of research progress on sustainability study of AM process is briefly reviewed. It is obvious that most current research focuses on the sustainability of the manufacturing process but neglects the impact from the design stage. The unique capabilities of the AM process may bring more freedom on the design stage, which may further improve the sustainability of a designed product and reduce its environmental impact during manufacturing stage. Thus, this chapter aims to provide a general sustainable design methodology for AM processes. To introduce this general design methodology, those AM-enabled design methods are first reviewed. Based on the existing AM-enabled design method, a general sustainable design methodology for

AM processes is proposed and discussed in detail. Finally, a brief case study is provided to illustrate and validate the proposed design methodology. Generally, the proposed design methodology can reduce the product's environmental impact during the manufacturing process by optimizing the design-dependent parameters which may cause the major environmental impact. Moreover, the parts count can also be reduced through functional integration and physical integration steps in the proposed design methodology. The reduction of the overall parts count definitely decreases assembly difficulties and further minimize the products' environmental impacts. It should be noted that the proposed design methodology also has certain limitations. For example, current design methodology only considers the environmental impact of the manufacturing process. However, sometimes, the environmental impact of products during other major life cycle phases may play an even more important role. For example, the weight of an aircraft may not only affect its environmental impact during the manufacturing phase but also has a great effect on its service phase. The lower the part weight the less fuel it uses. Thus, this proposed design methodology needs to be extended to the full product life cycle in the future.

References

Adam GAO, Zimmer D (2014) Design for additive manufacturing—element transitions and aggregated structures. CIRP J Manuf Sci Technol 7:20–28

A. Engineering (2009) Version10.0, Users manual Optistruct., Inc., Troy, MI

Albers A, Matthiesen S, Ohmer M (2003) An innovative new basic model in design methodology for analysis and synthesis of technical systems. In: DS 31: Proceedings of ICED 03, the 14th International Conference on Engineering Design, Stockholm

Allaire G, Jouve F, Toader A-M (2002) A level-set method for shape optimization. CR Math 334:1125–1130

Andreasen MM, Kähler S, Lund T (1983) Design for assembly. Ifs Publications, London, UK

Ashby MF, Cebon D (1993) Materials selection in mechanical design. Le Journal de Physique IV 3:C7-1–C7-9

ASTM Standard (2012) WK38342 New Guide for Design for Additive Manufacturing_in draft. West Conshohocken, PA. http://www.astm.org/DATABASE.CART/WORKITEMS/WK38342.htm

Baumers M, Tuck C, Bourell D, Sreenivasan R, Hague R (2011) Sustainability of additive manufacturing: measuring the energy consumption of the laser sintering process. Proceed Inst Mech Eng, Part B: J Eng Manuf 225:2228–2239

Becker R, Grzesiak A, Henning A (2005) Rethink assembly design. Assembly Autom 25:262–266

Bendsøe MP, Kikuchi N (1988) Generating optimal topologies in structural design using a homogenization method. Comput Methods Appl Mech Eng 71:197–224

Bendsøe MP, Ben-Tal A, Zowe J (1994) Optimization methods for truss geometry and topology design. Struct Optim 7:141–159

Bin Maidin S (2011) Development of a design feature database to support design for additive manufacturing (DfAM), Dissertation, Loughborough University, UK

Blouin VY, Oschwald M, Hu Y, Fadel GM (2005) Design of functionally graded structures for enhanced thermal behavior. In: ASME 2005 International Design Engineering Technical Conferences and Computers and Information in Engineering Conference, 2005, pp 835–843

Boothroyd G, Dewhurst P, Knight WA, Press C (2002) Product design for manufacture and assembly. M. Dekker New York, USA

Bourhis FL, Kerbrat O, Dembinski L, Hascoet J-Y, Mognol P (2014) Predictive model for environmental assessment in additive manufacturing process. CIRP Procedia 15:26–31

Boyard N, Rivette M, Christmann O, Richir S (2014) A design methodology for parts using additive manufacturing, high value manufacturing: advanced research in virtual and rapid prototyping, pp 399–404

Bralia J (1986) Handbook of product design for manufacturing: a practical guide to low-cost production. McGraw-Hill Book Company 1986:1120

Byun HS, Lee KH (2006) Determination of the optimal build direction for different rapid prototyping processes using multi-criterion decision making. Robot Comput-Integr Manuf 22:69–80

Castilho M, Dias M, Gbureck U, Groll J, Fernandes P, Pires I et al (2013) Fabrication of computationally designed scaffolds by low temperature 3D printing. Biofabrication 5:035012

Chen Z, Gao L, Qiu H, Shao X (2009) Combining genetic algorithms with optimality criteria method for topology optimization. In: Fourth International Conference on Bio-Inspired Computing, 2009. BIC-TA'09, pp 1–6

Chen Y, Zhou S, Li Q (2011) Microstructure design of biodegradable scaffold and its effect on tissue regeneration. Biomaterials 32:5003–5014

Delgado J, Ciurana J (2012) Mechanical characterisation of metal material properties in additive layer processes. Int J Mechatron Manuf Syst 5:189–213

Dorn WS, Gomory RE, Greenberg HJ (1964) Automatic design of optimal structures. J De Mec 3:25–52

Evans A, Hutchinson J, Fleck N, Ashby M, Wadley H (2001) The topological design of multifunctional cellular metals. Prog Mater Sci 46:309–327

E. Inc (2014) Data sheet of 316 stainless steel infiltrated with bronze, 13 Mar 2014

Faludi J, Bayley C, Bhogal S, Iribarne M (2015) Comparing environmental impacts of additive manufacturing vs traditional machining via life-cycle assessment. Rapid Prototyping J 21:14–33

Faur C, Crainic N, Sticlaru C, Oancea C (2013) Rapid prototyping technique in the preoperative planning for total hip arthroplasty with custom femoral components. Wien Klin Wochenschr 125:144–149

Gibson I, Rosen DW, Stucker B (2010) Additive manufacturing technologies rapid prototyping to direct digital manufacturing. Available: http://site.ebrary.com/id/10356040

Goedkoop M, Heijungs R, Huijbregts M, De Schryver A, Struijs J, van Zelm R (2008) ReCiPe 2008, a life cycle impact assessment method which comprises harmonised category indicators at the midpoint and the endpoint level, vol 1, 2009

Gu P, Hashemian M, Nee A (2004) Adaptable design. CIRP Ann-Manuf Technol 53:539–557

Harper SR, Thurston DL (2008) Incorporating environmental impacts in strategic redesign of an engineered system. J Mech Des 130:031101

Hopkinson N, Hague RJM, Dickens PM (2006) Rapid manufacturing : an industrial revolution for the digital age. Wiley, Chichester, England

Huang R, Riddle M, Graziano D, Warren J, Das S, Nimbalkar S et al (2015) Energy and emissions saving potential of additive manufacturing: the case of lightweight aircraft components. J Cleaner Prod

Kellens K, Dewulf W, Deprez W, Yasa E, Duflou J (2010) Environmental analysis of SLM and SLS manufacturing processes. In: Proceedings of LCE2010 Conference, pp 423–428

Kellens K, Yasa E, Renaldi R, Dewulf W, Kruth JP, Duflou JR (2011) Energy and resource efficiency of SLS/SLM processes. In: 22nd Annual International Solid Freeform Fabrication Symposium—An Additive Manufacturing Conference, SFF 2011, pp 1–16

Kellens K, Dewulf W, Overcash M, Hauschild M, Duflou J (2012) Methodology for systematic analysis and improvement of manufacturing unit process life-cycle inventory (UPLCI)—CO2PE! initiative (cooperative effort on process emissions in manufacturing). Part 1: Methodology description. Int J Life Cycle Assess 17:69–78

Kim GD, Oh YT (2008) A benchmark study on rapid prototyping processes and machines: quantitative comparisons of mechanical properties, accuracy, roughness, speed, and material cost. 222:201–15

Kreiger M, Pearce JM (2013) Environmental impacts of distributed manufacturing from 3-D printing of polymer components and products. In: MRS Proceedings, pp 85–90

Kruf W, van de Vorst B, Maalderink H, Kamperman N (2001) Design for rapid manufacturing Functional SLS Parts. In: Intelligent Production Machines and Systems-2nd I* PROMS Virtual International Conference 2011, p 389

Kumar A, Choudhary S (2015) Analysis and substitution of metal triple tree/yoke of motorcycle with plastic triple tree. HCTL Open Int J Technol Innovations Res (IJTIR) 16

Le Bourhis F, Kerbrat O, Hascoet JY, Mognol P (2013) Sustainable manufacturing: evaluation and modeling of environmental impacts in additive manufacturing. Int J Adv Manuf Technol 69:1927–1939

Lopes AJ, MacDonald E, Wicker RB (2012) Integrating stereolithography and direct print technologies for 3D structural electronics fabrication. Rapid Prototyping J 18:129–143

Luo Y, Ji Z, Leu MC, Caudill R (1999) Environmental performance analysis of solid freedom fabrication processes. In: Proceedings of the 1999 IEEE International Symposium on Electronics and the Environment, 1999. ISEE-1999, pp. 1–6

Ma Z-D, Wang H, Kikuchi N, Pierre C, Raju B (2006) Experimental validation and prototyping of optimum designs obtained from topology optimization. Struct Multi Optim 31:333–343

Mahesh M, Wong Y, Fuh J, Loh H (2004) Benchmarking for comparative evaluation of RP systems and processes. Rapid Prototyping J 10:123–135

Maidin SB, Campbell I, Pei E (2012) Development of a design feature database to support design for additive manufacturing. Assembly Autom 32:235–244

Mani M, Lyons KW, Gupta SK (2014) Sustainability characterization for additive manufacturing. J Res Nat Inst Stand Technol 119:419–428

Mavroidis C, DeLaurentis KJ, Won J, Alam M (2001) Fabrication of non-assembly mechanisms and robotic systems using rapid prototyping. J Mech Des 123:516–524

Meteyer S, Xu X, Perry N, Zhao YF (2014) Energy and material flow analysis of binder-jetting additive manufacturing processes. Procedia CIRP 15:19–25

Mognol P, Lepicart D, Perry N (2006) Rapid prototyping: energy and environment in the spotlight. Rapid Prototyping J 12:26–34

Morrow WR, Qi H, Kim I, Mazumder J, Skerlos SJ (2007) Environmental aspects of laser-based and conventional tool and die manufacturing. J Cleaner Prod 15:932–943

Murr LE, Gaytan SM, Medina F, Lopez H, Martinez E, MacHado BI et al (2010) Next-generation biomedical implants using additive manufacturing of complex cellular and functional mesh arrays. Philos Trans Royal Soc A: Math, Phys Eng Sci 368:1999–2032

Otto KN, Wood KL (1998) Product evolution: a reverse engineering and redesign methodology. Res Eng Des 10:226–243

Otto KN, Wood KL (2001) Product design: techniques in reverse engineering and new product development. Prentice Hall, Upper Saddle River, NJ

Oxman N, Keating S, Tsai E (2012) Functionally graded rapid prototyping. In: 5th International Conference on Advanced Research in Virtual and Physical Prototyping, VR@P 2011, 28 Sept 2011–1 Oct 2011, Leiria, Portugal, pp 483–489

Pahl G, Beitz W, Feldhusen J, Grote K-H (2007) Engineering design: a systematic approach, vol 157. Springer, US

Perez KB, Williams CB (2013) Combining additive manufacturing and direct write for integrated electronics—A review. In: 24th International Solid Freeform Fabrication Symposium—An Additive Manufacturing Conference, SFF 2013, 12–14 Aug 2013, Austin, TX, United states, pp 962–979

Popsecu D (2007) Design for rapid prototyping: implementation of design rules regarding the form and dimensional accuracy of rp *prototypes*

Rännar L-E, Glad A, Gustafson C-G (2007) Efficient cooling with tool inserts manufactured by electron beam melting. Rapid Prototyping J 13:128–135

Regenfuss P, Ebert R, Exner H (2007) Laser micro sintering–a versatile Instrument for the generation of microparts. Laser Tech J 4:26–31

Reich Y (1995) A critical review of general design theory. Res Eng Des 7:1–18

Rodrigue H, Rivette M, Calatoru V, Richir S (2011) Une méthodologie de conception pour la fabrication additive

Rosen DW (2007a) Computer-aided design for additive manufacturing of cellular structures. Comput-Aided Des Appl 4:585–594

Rosen DW (2007) Design for additive manufacturing: a method to explore unexplored regions of the design space. In: Eighteenth Annual Solid Freeform Fabrication Symposium, pp 402–415

Rozvany G, Zhou M, Birker T (1992) Generalized shape optimization without homogenization. Struct Optim 4:250–252

Segonds F (2011) Contribution to the integration of a collaborative design environment in the early stages of design. PhD, Arts et Metiers ParisTech

Seliger G, Khraisheh MK, Jawahir IS (eds) (2011) Advances in sustainable manufacturing. In: Proceedings of the 8th Global Conference on Sustainable Manufacturing, Berlin, Heidelberg: Springer.

Shellabear M (1999) Benchmark study of accuracy and surface quality in RP models, Brite/EuRam Report BE-2051, Task, 4

Singhal SK, Jain PK, Pandey PM, Nagpal AK (2009) Optimum part deposition orientation for multiple objectives in SL and SLS prototyping. Int J Prod Res 47:6375–6396

Sreenivasan R, Goel A, Bourell DL (2010) Sustainability issues in laser-based additive manufacturing. In: Physics Procedia, pp 81–90

Suh NP (1990) The principles of design. Oxford University Press, New York, USA

Suh NP (1998) Axiomatic design theory for systems. Res Eng Design 10:189–209

Tang Y, Hascoet JV, Zhao YF (2014) Integration of topological and functional optimization in design for additive manufacturing, presented at the ASME 2014 12th Biennial Conference on Engineering Systems Copenhagen, Denmark

Telenko C, Conner Seepersad C (2012) A comparison of the energy efficiency of selective laser sintering and injection molding of nylon parts. Rapid Prototyping J 18:472–481

Thomas D (2010) The development of design rules for selective laser melting. Dissertation, University of Wales, UK

Tomiyama T (2006) A classification of design theories and methodologies. In: ASME 2006 International Design Engineering Technical Conferences and Computers and Information in Engineering Conference, 2006, pp. 43–51

Tomiyama T, Gu P, Jin Y, Lutters D, Kind C, Kimura F (2009) Design methodologies: industrial and educational applications. CIRP Ann-Manuf Technol 58:543–565

Vayre B, Vignat F, Villeneuve F (2012) Designing for additive manufacturing. Procedia CIRP 3:632–637

Vayre B, Vignat F, Villeneuve F (2013) Identification on some design key parameters for additive manufacturing: application on electron beam melting. Forty Sixth Cirp Conference on Manufacturing Systems 2013, vol 7, pp 264–269

Wang SY, Tai K (2005) Structural topology design optimization using genetic algorithms with a bit-array representation. Comput Methods Appl Mech Eng 194:3749–3770

Wang MY, Wang X, Guo D (2003) A level set method for structural topology optimization. Comput Methods Appl Mech Eng 192:227–246

Watts D, Hague R (2006) Exploiting the design freedom of RM. In: Proceeding of the Solid Freeform Fabrication Symp., Austin, TX, 14–16 Aug 2006, pp 656–67

Weber C (2005) CPM/PDD–an extended theoretical approach to modelling products and product development processes. In: Proceedings of the 2nd German-Israeli Symposium on Advances in Methods and Systems for Development of Products and Processes, 2005, pp 159–179

Weidema B, Bauer C, Hischier R, Mutel C, Nemecek T, Vadenbo C et al (2011) Overview and methodology. Data Qual Guidel Ecoinvent Database Version 3:76–84

Wilson JM, Piya C, Shin YC, Zhao F, Ramani K (2014) Remanufacturing of turbine blades by laser direct deposition with its energy and environmental impact analysis. J Cleaner Prod 80:170–178

Wohlers TT (2010) Wohlers Report 2010: Additive Manufacturing State of the Inudstry: Annual Worldwide Progress Report: Wohlers Associates

www.3Ders.org (2013, July 20). 3D printing can cut material consumption by 75%, CO_2 emissions by 40%. Available: http://www.3ders.org/articles/20131024-3d-printing-can-cut-material-consumption-co2-emissions.html

Xie Y, Steven GP (1993) A simple evolutionary procedure for structural optimization. Comput Struct 49:885–896

Xu X, Meteyer S, Perry N, Zhao YF (2014) Energy consumption model of binder-jetting additive manufacturing processes. Int J Prod Res 1–11

Yang S, Zhao YF (2015) Additive manufacturing-enabled design theory and methodology: a critical review. Int J Adv Manuf Technol 1–16

Yang S, Zhao Y (2015) Additive manufacturing-enabled design theory and methodology: a critical review. Int J Adv Manuf Technol 1–16

Yang S, Tang Y, Zhao YF (2015) A new part consolidation method to embrace the design freedom of additive manufacturing. J Manuf Proc

Yoon HS, Lee JY, Kim HS, Kim MS, Kim ES, Shin YJ et al (2014) A comparison of energy consumption in bulk forming, subtractive, and additive processes: review and case study. Int J Precis Eng Manuf—Green Technol 1:261–279

Young V, Querin O, Steven G, Xie Y (1999) 3D and multiple load case bi-directional evolutionary structural optimization (BESO). Struct Optim 18:183–192

Zhou M, Xi J, Yan J (2004) Modeling and processing of functionally graded materials for rapid prototyping. J Mater Process Technol 146:396–402

Redesigning Production Systems

Jennifer Loy and Peter Tatham

Abstract If it was possible to wind back the clock on the first Industrial Revolution, then a redesign of production systems, based on the information available now, would focus on reducing environmental impacts, maximising resources and adding value to all products created, as well as taking into account the health and wellbeing of workers and the distribution of populations. Additive manufacturing, combined with digital communication technologies, delivers the possibility that many of the goals can be achieved—leading to a much healthier planet. Based on current research into sustainability and additive manufacturing outcomes, this chapter provides a vision for the redesign of current production systems, supply chains and values that serves as starting point for re-establishing the human relationship with manufacturing and business practice. Current drivers for change are discussed and opportunities for reducing the environmental impact of production systems directly enabled by additive manufacturing are then considered. These are based on integrating additive manufacturing into the supply chain and the potential impact on the development cycle, inventory management, logistic postponement and the management of spare parts.

Keywords 3D printing · Additive manufacturing · Global connectivity · Logistics · Supply chain · Sustainability

1 Introduction: Systems in Crisis

Looking back, the Industrial Revolution was not a particularly good idea. Whilst there are obvious dangers in romanticizing the pre-industrial era, there is growing evidence that the reality of the Industrial Revolution as it transpired was the

J. Loy (✉) · P. Tatham
QCA and Griffith School of Engineering, Griffith University, Southport, Gold Coast,
QLD 4222, Australia
e-mail: j.loy@griffith.edu.au

© Springer Science+Business Media Singapore 2016   145
S.S. Muthu and M.M. Savalani (eds.), *Handbook of Sustainability
in Additive Manufacturing*, Environmental Footprints and Eco-design
of Products and Processes, DOI 10.1007/978-981-10-0549-7_7

thin end of the wedge in terms of negatively changing the longer-term human relationship with its environment. Indeed, arguably, since the end of the Industrial Revolution humans have been on a slippery slope towards self-destruction that, because of the over-consumption of resources, mimics the demise of a myriad of species throughout nature over the history of the planet. Individual human civilizations have also, in many instances, risen to incredible heights of sophistication and complexity where longevity must have seemed assured, before falling to shadows in the sand of ancient cities and decaying monuments.

To consider the health of the planet and the need for a societal group to operate in a particular niche within the limited provision of the planet as a whole, the impact of humans' self-destructive behaviour has, aided by technology, dramatically increased over the last 200 years. A key example is the creation of waste that cannot be reclaimed as illustrated in the growing problems of discarded plastic in the Great Pacific Garbage Patch (Moore 2009) and technological debris in the stratosphere. Industrialization has not only provided the means to create more consumer waste; the resulting urbanization as a consequence of the changing forms of industrialization over the last 200 years has created larger and larger cities that change the human relationship with the environment. In 2014 more than 50 % of the population lived in cities for the first time. It is estimated that by 2025 there will be 30 cities with a population more than 10 million. These cities have massively invested infrastructures, thereby reducing the ability of populations to provide an agile response to changing production demands. The social consequences of this pattern of urbanized industrialization includes those associated with unemployment when a significant industry fails, as well as the deterioration of the cities themselves, as can be seen in Detroit in the aftermath of the collapse of the previously dominant car industry.

Manufacturing in the nineteenth and early twentieth centuries led to increased consumerism which, in turn, generated increased employment. Products cost relatively less than previously, shifting consumerism from 'needs' to 'wants' (Forty 1992). However, even with the increased ability to produce more goods and to ship those goods around world, the prevailing attitude towards consumerism maintained some restraint in Western societies, until the aftermath of the Second World War created an economic downturn. This post-war slump led to a drive to stimulate the economy through increased consumerism. Eisenhower's exhortation to Americans to 'Buy, buy anything' to help the economy was reflected in changing marketing tactics, such as the introduction of new car models each year to stimulate trade, and built-in obsolescence with the growth of a 'throw away' society. Vance Packard was an early voice in raising concerns about such practices in his trilogy of books, led in 1960 by *The Waste Makers* (Packard 2011), which criticized the evolving systems of production, consumption and waste-making. For industrial designers, Papanek's seminal work (Papanek 2005) questioned the actions and responsibilities of designers in such a globalized production environment. Up until the mid-1980s there was an eco-design agenda that ran alongside commercial production practice. However, by the late 1990s, this relatively small fringe movement was superseded by a mainstream shift in thinking and practice brought about by a growing understanding of the sustainability imperative.

Industrial Design was initially created as a discipline in response to a need by manufacturers for new workers who could combine the documentation of a products production sequence with the ability to understand and to respond aesthetically and functionally in their design work to the needs, desires and aspirations of specific target markets. For professional Industrial Designers in the last 20 years of the twentieth century, the ability to take eco-design issues into account was predominantly determined by the client who may/may not give permission to proceed in such a direction. Thus, it has only been the broader political agenda of the last 10 years that has influenced clients—usually for direct or indirect economic reasons—to accept products that demonstrated an eco-design approach. Indeed, as the Bruntland Commission's definition of sustainability began to be more widely influential, the rhetoric changed. Essentially there was a shift in thinking in relation to the impact of production processes on the environment. This contributed to the development of life cycle assessments and a move towards triple bottom line accounting—as outlined in *Cannibals with Forks: Triple Bottom Line of 21st Century Business* (Elkington 1999).

The resultant economic incentive of creating products more aligned to the thinking at this time created a shift in the nature of eco-design products from the equivalent of 'vegan to vegetarian', and the consequential move into more mainstream production. This led to the eco-pluralistic approach described by Fuad-Luke in his book, *Ecodesign: The Source Book* (2006), and a proliferation of works from around the globe aimed at addressing sustainability imperatives. For example Fuad-Luke (2005, p. 15) highlighted the growing number of companies seeking membership on the Dow Jones Sustainability indexes and he argued that triple bottom line accounting was creating an opportunity for businesses to work with designers to slow environmental degradation through the introduction of more environmentally benign products. Fuad-Luke saw this as a 'win–win' for businesses and the environment, with manufacturers spending less on raw materials and production, whilst creating products that were better for the environment, more efficient and better value. Fuad-Luke also argued that governments would then be able to reduce spending on regulatory enforcement and all concerned would benefit from an enhanced quality of life and better product margins.

For the Industrial Designer, this ecopluralistic approach was a difficult one to sell to the client as it included a range of very different strategies. However Hawken and Lovins in *Natural Capitalism: The Next Industrial Revolution* (2005), unified the growing acceptance of the importance of sustainability drivers as integral to production practices as the 'next industrial revolution'. These authors argued that a positive approach to sustainability by manufacturers and industrial designers would lead to a new form of 'natural capitalism' where a novel industrial system would emerge that operated "as if living systems mattered" (Hawken and Lovins 2005, p. 9). Sadly, notwithstanding this positive positioning of industrial design and production, things have been complicated. For a start, the recent (2015) example of Volkswagen and the emissions scandal demonstrates that reducing regulatory enforcement is not an option where human greed and profit are involved. However, even with the purest of motives, the complications of working within current production systems to try and mitigate environmental impact are incredibly challenging. The work of McDonough and Braungart in *Cradle to Cradle* (2002) attacked the naiveté of many of the approaches to reducing environmental impact—such as the fundamental ideas of reuse and recycling—advocated by many authors and commentators. Rather, McDonough and Braungart argued that recycling simply postponed the inevitable as the product in question, ultimately, still goes on to become landfill, and, indeed, may do unanticipated additional damage to the environment on the way: "When you went shopping for a carpet recently, you deliberately chose one made from recycled polyester soda bottles. Recycled? Perhaps it would be more accurate to say downcycled. Good intentions aside, your rug is made of things that were never designed with this further use in mind, and wrestling them into this form has required as much energy—and generated as much waste—as producing a new carpet...moreover the recycling process may have introduced even more harmful additives than a conventional product contains, and it might be off-gassing and abrading then into your house at an even higher rate" (McDonough and Braungart 2002, p. 4).

As a result, McDonough and Braungart (2002, p. 163) called for a more complex rethinking of manufacturing and human systems, embodying a commitment to social equity, ecology and the economic bottom line thinking that was informed throughout the length of its supply chain by thorough research, suggesting that this would result in what they termed 'eco-efficiency'. McDonough and Braungart discussed the work of Ford in this area as an example of how an established organization of significant size could re-invent itself with an eco-efficiency agenda, arguing that it was "not possible (nor would it be necessarily desirable) to simply sweep away long-established methods of working, designing and decision-making" (McDonough and Braungart 2002, p. 165). Yet these authors also call for new thinking in addressing the "messy, burdensome and threatening, even overwhelming" challenges that established manufacturing practices face.

The maturing of sustainability thinking into a more comprehensive rethinking of the fundamentals of complex systems has led to strategies for increased eco-efficiency, including 'dematerialization' and the move towards product service systems (Ryan 2004, p. 52), which involves rethinking the function of a product in terms of the service it provides, and developing an outcome focussed on solutions, not products. Indeed, a similar concept of 'Service Dominant Logic' (SDL) has emerged in marketing literature (Vargo and Lusch 2004) and this has also migrated as a theme in emerging supply chain management thinking (Lusch and Vargo 2014).

However, contrary to the argument that McDonough and Braungart made about retaining long-established production systems and methods, it is clear that the changing digital environment, coupled with a shift in thinking to agile, product service system thinking, provides an alternative view of design and manufacturing which focuses on small businesses and new practices. Additive manufacturing, commonly known as 3D printing (3DP), is integral to this opportunity to rethink the relationship of humans to their product support system and to their environment. As proposed by Aldersey-Williams (2011) in *The New Tin Ear: Manufacturing, Materials and the Rise of the User-Maker*, new systems could now be developed that create commercial and production practices that return societies to pre-industrial revolution organizational approaches.

For example, in the last 5 years, the rise of crowd sourcing and online retailing has shown that, contrary to the focus of conventional thinking on the dominance of traditional industrial and business practices, agile production, the bourgeoning of 'start-ups' and digital innovative manufacturing offer new directions that could ultimately prove more significant, and therefore potentially more effective in combating the unsustainability of current behaviours.

2 The Change Is Now

Jeremy Leggett highlighted resourcing concerns about dwindling global energy supplies in his book *Half Gone* (2005). The focus around this time began to shift from improving eco-efficiencies to creating radical change. This aligned with the view that Leonard Mau and colleagues developed in *Massive Change* (Mau et al. 2004), where they argued that it was necessary to have a genuine impact on the failing infrastructures that were supporting what they considered to be the artificially maintained lifestyle of wealthy consumers around the world. They deemed it necessary for society to have a more acute awareness of "real life" (Mau et al. 2004, p. 6) and the "bewildering complexity of our increasingly interconnected (and designed) world" (Mau et al. 2004, p. 11). They also suggested a significant shift in thinking in terms of design economies and that the use of Product Service System thinking was increasingly appropriate: "instead of looking at product design, we looked at the economics of movement. Instead of isolating graphic design, we considered the economies of information, and so on. The patterns that emerged reveal complexity, integrated thinking across disciplines and unprecedented interconnectivity" (Mau et al. 2004, p. 16).

The *Massive Change* project instigated by *Institute Without Boundaries* attempted to move design outwards into the community with the idea of creating what it termed "advanced capitalism, advanced socialism, and advanced globalisation" (Mau et al. 2004, p. 19). The "future of global design is fundamentally collaborative" was a response to the analysis of Buckminster Fuller in the *World Resources Inventory*, and Mau et al. focus on his comments that "There are very few men today who are disciplined to comprehend the totally integrating significance of the 99 % invisible activity which is coalescing to reshape our future. There are approximately no warnings being given to society regarding the great changes ahead. There is only the ominous general apprehension that man may be about to annihilate himself" (Mau et al. 2004, p. 22).

Over the following decade, the need for a paradigm shift in relation to supply chains and sustainability was emphasized by the development of yet more complex understandings of the impact of conventional production and consumption systems on the health of the planet. Flannery, in his book *The Weather Makers* (2006), demanded action from individuals to reduce their individual consumption by 70 % in the face of damage that predominantly started in the 1950s as cars and household devices began to proliferate. He argued that the initial damage was inflicted through ignorance, but that when the knowledge of the impact on the environment of profligate behaviours became freely available, then it became a matter of individual responsibility. According to Flannery, one of the challenges in this is that the question of 'what constitutes dangerous climate change?' raises the additional question of: "dangerous to whom? For the Inuit in the Arctic a damaging threshold has already been crossed. Their primary food sources of caribou and seal are now difficult to find as a result of climate change and their villages are under threat" (Flannery 2007, p. 161). Thus, for the collective consciousness of people on one side of the planet to take responsibility for the impact of their actions on people on the other side of the planet, it requires the human population to start to think as a single organism.

In his book *The Future*, Gore (2014, p. 15) judged that there was "a clear consensus that the future now emerging is extremely different from anything we have ever known in the past. It is a difference not of degree but of kind". He argued for a paradigm shift—not an evolutionary change but rather a "massive global transformation of our energy, industrial, agricultural, and construction technologies in order to re-establish a healthy and balanced relationship between human civilisation and the future". He went on to identify a number of drivers for change based on a growing awareness of worldwide, rapid, unsustainable growth and its impact on resource consumption. Gore also highlighted several fundamental concerns including the depletion of topsoil, the pressures on freshwater supplies and the increase in pollution. He criticized current economic patterns and outputs as "measured and guided by an absurd and distorted set of universally accepted metrics that blinds us to the destructive consequences of the self-deceiving choice we are routinely making" (Gore 2013, p. 14).

3 Global Connectivity, Drivers for Change and Opportunities for Change

The growing sustainability imperative of the last 20 years has been accompanied by equally significant developments in digital technologies. The transition from mainframes to personal computers and thence to hand-held devices has led to changes in human behaviour enabled by computing technologies that were not anticipated. Arguably, the most significant of these has been the communications revolution. From its inception as an information repository, the Internet has matured with the development of Web 2.0, which refers to the interactive capacity of networked technologies—the Internet of things.

Social media websites, now extended through smartphone digital communication platforms such as Twitter, have provided the means for collective discussion and near-instantaneous information sharing and decision-making around the world. This changing communication environment is breaking down conventional divisions. It is an example of how unanticipated change can occur in a relatively short time, providing less of an evolution than a step-change. Made possible by innovations in technology, it creates a wealth of new possibilities when considering future scenarios.

This provides further credibility to the arguments made by Aldersey-Williams (2011) for changes in how society is organized by production, and the ideas of Mau et al. (2004) suggesting individuals could act as change agents. The development of a written language was described as a factor in the emergence of significant historical civilizations, such as Mesopotamia (Gore 2014, p. 50). It was considered responsible for reducing the individuals' ability to retain social knowledge, but also credited with increasing the ability of a social order to retain and enhance collective knowledge, resulting in increasingly complex understandings and behaviours. Improved communication tools were credited with contributing to the breakdown of the feudal system during the Agricultural Revolution (Gore 2014).

In a similar way, the spread of digital communication technologies is creating new ways of thinking, new societal structures, and new economic structures: "The transformation of the global economy is best understood as an emergent phenomenon—that is, one in which the whole is not only greater than the sum of its parts, but very different from the sum of its parts in important and powerful ways. It represents something new—not just a more interconnected collection of the same national and regional economies that used to interact with one another, but a completely new entity with different internal dynamics, patterns, momentum, and raw power than what we have been familiar with in the past" (Gore 2014, p. 5).

Mau et al. (2004, p. 219) point to a 'citizen revolution' of social entrepreneurs with ethics necessary to challenge "free-market fundamentalism and economic globalisation" emerging as 'changemakers' around the world. However, Gore argues that "the simultaneous deployment of the Internet and ubiquitous computing power have created a planet-wide extension of the human nervous system that transmits information, thoughts, and feelings to and from billions of people at the speed of light" (Gore 2014, p. 44), which suggests something quite different. The vast majority of digital information currently transmitted happens without any physical human interaction, creating what are termed 'Big Data' sets that are increasingly mined for information to drive decision-making external to human minds. Thus, Gore refers to the development of a 'Global Mind': "Our societies, culture, politics, commerce, educational systems, ways of relating to one another—and our ways of thinking—are all being profoundly reorganized with the emergence of the Global Mind and the growth of digital information at exponential rates" (Gore 2014, p. 44).

Rather than working with existing systems of production and consumption, Mau et al. argued to "seamlessly integrate all supply and demand around the world" (2004, p. 126), describing the notion of the 'intermodal' and the resultant global infrastructure as the "accidental avant-garde of a new global politics of ecology". This proposition is even more realistic than before if the digital revolution of the last 20 years is considered and has included the development of digital fabrication technologies that have evolved 3DP from a prototyping technology to one suitable for direct manufacturing as well as the global connectivity needed for distributed manufacturing.

The digital communication revolution that has occurred since the spread of the Internet over the past 20 years has created unexpected change and opportunities to revise completely relationships between producers and consumers in new markets: "National policies, regional strategies, and long accepted economic theories are now irrelevant to the new realities of our new hyper-connected, tightly integrated, highly interactive, and technologically revolutionized economy" (Gore 2014, p. 4). If humans are to confront sustainability issues with a sense of worldwide, collective responsibility, and react as a single organism for the benefit of the whole, rather than being driven by the short-term priorities of individuals, or relatively small economic or cultural groups, then the emergence of a system for collective thought and decision-making is needed. The challenges and opportunities to address all aspects of sustainability—from manufacturing, to sociocultural sustainability—are now different to those encountered any time previously. This, in turn, requires responses based on emerging thinking that are informed by our knowledge and understanding of new technologies that, in turn, offer new strategic opportunities to be seized leveraged.

4 Additive Manufacturing and Global Connectivity for Sustainable Design and Production

The digital revolution of recent years has seen the development of multiple means to communicate, generate and manipulate data relevant to manufacturing. These have included scanning technologies and point cloud manipulation software as the basis for 3D modelling. The evolving range of technologies has created a myriad of opportunities for service-based industries. However, for physical products, the most revolutionary digital technology developed has been 3D printing.

Since its initial development as a prototyping technology that started with stereolithography, 3DP (as it is now known) has expanded to encompass a broad range of technologies based on additive rather than subtractive building. These include fused deposition modelling, selective laser sintering and direct laser melting. Key to all these technologies is that the products built do not rely on pre-formed moulds. This means they can be distributed individually and printed on demand from digital files (STLs). In addition, the fact that the models are created on the computer and printed without the investment needed for a mould means that it is possible to customize production so that each print is different and bespoke. This shatters conventional supply chain practices for manufacturing and requires a rethink of the organization of production enabled by 3DP. This has clear implications for sustainability based on our developing understanding of the step-change required to meet the environmental challenges impacting on manufacturing and consumerism.

5 Supply Chain Management (SCM) Implications of 3D Printing

The basic challenge for a supply chain manager in developing the organization of production and distribution of a product is that of achieving what are known as the '5 Rights' of supply chain management—the *right* product at the *right* place at the *right* time in the *right* quantity and quality and at the *right* cost—or, to put it more succinctly, ensuring that supply = demand as accurately as possible. 3DP impacts on the levers of this equation (supply, demand, and cost) in multiple ways, including the Development Cycle, Inventory Management, Logistic Postponement, and Management of Spare Parts. In doing so it also offers an example of a pilot case study illustrating the game-changing nature of the impact of 3DP on the supply chain in the context of a developing country or one affected by the impact of a disaster or complex emergency (Loy et al. 2015; Tatham et al. 2015).

5.1 The Development Cycle

There is broad recognition within the literature that, globally, businesses are experiencing a period of growing turbulence. Thus, for example, Doheny et al. (2012, p. 2) suggest that the potential for catastrophic failure in a company's supply chain has: "… become more acute in recent years as rising volatility, uncertainty, and business complexity have made reacting to—and planning for—changing market conditions more difficult than ever." A similar perspective is found within both the academic literature (Singh 2009; Christopher and Holweg 2011) and in the practitioner focussed discussions such as the 'Shell Energy Scenarios to 2050' that envisage "an era of revolutionary transitions and considerable turbulence" (Shell 2008, p. 10). Within this overall context, change brought about by the rapid expansion of digital technologies is a significant factor in creating this business turbulence. Changing customer relations, digital fabrication technologies—in particular 3DP—and 'Big Data' are altering the demand for products and their nature.

In parallel, as noted by Christopher (2011), forecast error—that is the difference between anticipated demand and the actual usage—increases more than proportionately over time. Because of this, any strategy to reduce the lead time gap between procurement and product delivery is a clear candidate for serious consideration. If successful, this would result in the reduction of 'just-in-case' inventory or lost sales that result from a product being out of stock when the anticipated demand falls short of the actual demand. For the supply chain manager, the potential of the use of 3DP would be to mitigate these problems.

At the new product development level, the inherent flexibility of 3DP and the relatively low cost of creating test pieces using the technology have led to reduced prototyping timescales being reported by major manufacturers. Thus, the Ford Motor Company has provided an example of this reduction in practice, where they reported a reduction in the average time of their prototyping development for parts from 3–4 months to a matter of weeks (Shinall 2013). Furthermore, not only is the prototyping process swifter, it is also possible to undertake multiple repetitions of the process and modified iterations of the product which, in turn, have the potential to lead to more innovative and/or higher quality products.

In addition to the above benefits which reflect efficiency improvements in support of the creation of mass-produced components, 3DP also has the capability to create bespoke items ab initio rather than relying on the downstream adjustment of a standard item to meet the specific requirements of a particular context. It should also be noted that, in some cases, the use of 3DP actually allows the creation of a component that would otherwise be extremely challenging using conventional means—an example being in-line filtration within a piece of pipework.

5.2 Inventory Management: Production and Distribution

It is in relation to the management of inventory, both in terms of production and distribution, that 3DP has significant potential to provide greater logistical efficiency and effectiveness. The key underpinning concept of 3DP is that it is an additive process. On a like for like basis, it is highly likely that the mass and volume of the raw material required to produce a given item are less than in traditional mass production techniques which are subtractive in nature. This is true even if one takes into account the support material required during printing—although, hopefully, the product would have been designed to minimize the support filament in a fused deposition modeling process, or designed to nest to reduce the amount of support powder in a selective laser sintering process that would need to be down-cycled.

Put simply, if the product is designed for 3DP, then almost all the raw material input into the production process is converted into output. Indeed, in extreme cases, manufacturing products using traditional subtractive techniques such as milling can result in wastage of as much as 95 % of the original source material (Economist 2013). Thus, unsurprisingly, converting to 3DP has been estimated to result in approximate material costs savings per item of up to 25 % (Pannett 2014). This also reduces the cost and embodied energy impact in transporting raw materials to the production plant, together with a similar saving in local warehousing requirements.

Similarly, if not even more importantly, benefits pertain to the post-production process where, under the traditional model, finished products must be transported to the point of sale. Thus, under a 3DP distributed manufacturing model with the customization of products specific to site requirements, the point of production is significantly closer to the point of sale and this has a number of major logistic implications. First, the raw materials, typically, have a high mass:volume ratio. This means that they are more easily transported and at less cost. In addition, unlike finished goods, the packaging requirements are minimized. This is also important as pre-consumer packaging is a significant sustainability issue, so any reduction of the transportation of packaging is a clear benefit both in terms of avoiding nugatory production of these packaging materials also through a reduction in the volume required during transit. This combination of factors means that, in effect, unlike the current model, with 3DP there is less transportation of air around the globe, with clear financial and sustainability benefits. In addition, it is likely that there are reduced losses caused by damage and also what is euphemistically referred to within the industry as 'shrinkage' (i.e. theft).

The potential savings are significant on many fronts. For example, under the current production model in which multiple items are made in, say, China and shipped in their final form to the Americas, Europe or Australasia, the result is the existence of a 'stockpile' of finished goods that are in transit for a significant period of time—typically 2–3 months. Furthermore, as discussed above, because of the packaging and size/shape of the finished goods, much of this load may actually be air. In any event, this inventory has to be financed with a clear challenge to the cash-to-cash cycle implicit in the lead time between the factory gate and the retail outlet.

5.3 Logistic Postponement

Perhaps even more importantly, however, the existence of a lead time which is typically counted in months means that it is extremely difficult for a the supply chain to react to any short-notice positive or negative changes to the market described earlier. This highlights what is probably the most important benefit of 3DP as seen from a logistic perspective—namely that it comes close to the ultimate aim in achieving the concept of postponement. Put simply, the idea of postponement is that the decision to make a product in its final form is delayed for as long as possible, ideally until the demand for that product is clear. This approach has been adopted in many industries including, for example, clothing where manufacturers are increasingly producing items as 'vanilla' products. Relatively small quantities of these items can then be dyed in a variety of colors and tested in the market. Based on the resultant consumer feedback of the likely demand, more items can then be produced in the favored colors. In essence, this process allows the mass production of the core item, but at the same time delays the more expensive but value adding elements until there is a clearer understanding of the likely level and location of demand.

The use of 3DP has the potential to take this approach one stage further if implemented via, for example, a high street 'print shop' which would only make the required component when requested by the customer. In essence, the production process would wait until the demand has crystallized, thereby making this the near ultimate in postponement. Indeed, this approach can clearly be taken to the final step through the use of home printers, sales of which are forecast to double every year for at least the next three (Gartner 2014).

Putting to one side the potential for the ultimate in terms of local production that might be seen as a return to the pre-industrial age business model, the widespread development and use of 3DP has the potential to change the generic logistic model. This has, certainly over the last two decades, seen the closure of local distribution depots and the growth of large regional distribution centers (RDCs) in many countries. Such RDCs work on the basis of economies of scale and rely on both large volumes of goods shipped in and multiple deliveries to retail outlets. This has, as an example, been taken to a significant level in a large but relatively sparsely populated country, Australia, where the distance between RDC and a retail outlet may be in excess of 1800 km, with the clear challenges evident for achieving the '5 Rights' introduced earlier. Thus, the arguments for local production become significantly more attractive, especially when multiple items can be created from a single source material as in the case of 3DP. Indeed, to take this approach one stage further, one could easily envisage an extension of the Vendor Managed Inventory model whereby oversight of a range of items is undertaken locally.

5.4 Management of Spare Parts

Much of the above discussion relates to the potential benefits for improved efficiency in the creation of new items through the use of 3DP. However, this technology also has the potential to become even more influential in the case of the provision of spare parts. Here, the demand pattern for a given item is even less predictable than that for a whole product and this, in turn, magnifies the challenge of deciding which spares should be warehoused, in what quantities, and where. The introduction of 3DP does much to mitigate these challenges. The production of a component 'to order' clearly minimizes the warehousing requirements. In reality, it may not be appropriate to create a part only when it is required by the consumer, rather the demand may be satisfied from an item held on the warehouse shelf with its replacement being created via local 3DP. In any event, the key change here relates to the reduction of the uncertainty in demand and the associated forecasting challenge.

Extending this production of a complete item, the process can be taken one stage further where the actual printing takes place at the level of the consumer. Thus, for example, in a case outlined by Fawcett and Waller (2014), rather than spend significant time searching car salvage yards for spares for his collection of old vehicles, an individual was able to provide the necessary components with the minimum of difficulty through the combination of 3D scanning the defective part and 3D printing a replacement. Furthermore, this approach clearly lends itself to the modification of standard parts in order to meet a particular context. Thus, for example, an item might normally be constructed with a 90° bend, but if the particular context requires the use of 60° or 120°, then 3DP allows the creation of a modified component without the loss of structural integrity. Such an approach might be taken still further from a reactive model to one that is embedded in a continuous improvement cycle.

The significance of this potential to eliminate the manufacture of products that are only produced as spare parts and then *never actually required* is immense from an environmental impact point of view. The value in the raw materials is retained, the energy embodied in production is saved, the 'transport miles' are avoided, and the space in landfill negated. All of these benefits contribute to closing the loop for production, aligning with current sustainability strategies.

6 Humanitarian Logistic Case Study

The discussion to date has, unsurprisingly, focussed on the impact of 3DP in the context of a developed economy but, arguably, it has even greater potential to achieve logistic benefits in the context of a developing region and/or one that has been affected by a disaster or complex emergency. The core logistic benefits of 3DP (responding swiftly to an unanticipated demand, a reduced requirement for warehousing, reduced and more efficient transportation, the ability to modify a component to meet a particular need) are clearly applicable in the development/ disaster response context.

As an example, during recent field-based research in Kenya supported by the Humanitarian Innovation Fund, it was noted that a time lapse of 1–3 months between the raising of a demand and its fulfilment was not unusual (Tatham et al. 2015). In a similar way, Durgavich (2009) suggests that delays of up to 6 months for some countries have been encountered in achieving customs clearance. Self-evidently, therefore, the ability to produce a component locally 'on demand' has major potential to mitigate such delays. In addition, and in relation to warehousing, the minimising of the requirement to hold spare parts, etc. 'just-in-case' (i.e. the items may never actually be needed), has the potential to save significant costs, of particular importance in the humanitarian field.

This approach has dual benefits as the particular operational circumstances of remote regions are likely to require the ad hoc modification of items to meet the specific on-site requirements. Anticipating specific requirements, particularly prior to a location being operationally established, is difficult. The ability to modify products on-site prior to 3D printing could not only save replicating a product but also potentially ensure the continuation of a service that might otherwise have been delayed, whilst a replacement part not possible to modify by conventional means was sent from the RDC.

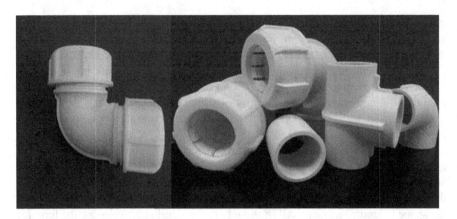

Fig. 1 Examples of Griffith University WASH project 3D printed connectors

These theoretical benefits were tested as part of pilot research hosted by Oxfam GB in Nairobi that was aimed at confirming the reality of the potential of 3DP on the ground. The focus of the research was the Water, Sanitation and Hygiene (WASH) project as such facilities are required in both development and disaster response operations in remote regions. WASH equipment is relatively low technology and does not operate at high pressure, with the result that any component failures were unlikely to be injurious or life-threatening. 3DP was used on-site to supplement the usual supply chain of piping and associated products. Specifications for particular products were gathered on-site, and then the information send via the Internet to Griffith University (in Queensland, Australia), where the design and testing of the product suitable for 3DP was developed. The file was then sent back to the site, where it was printed and tested. This approach was found to be of particular value in providing stop-gap products to ensure the flow of water was maintained or where a connector or piece of angled piping with an unusual geometry was required (Fig. 1).

From a sustainability point of view, the broader implications of a shift in production approach from overseas manufacturing and subsequent importation has clear benefits. The current model involves the management of an often administratively convoluted supply chain (because of, for example, war-torn provinces requiring products to be diverted), to one in which manufacturing on-site takes place. This approach has the potential to provide a blueprint for locally produced, bespoke products that could greatly reduce the environmental impact of providing the services required.

There are, inevitably, a number of issues needing to be addressed such as the existence and stability of a power supply. Currently, portable power generators and/or battery packs are available, but they are unlikely to have been optimized to minimize their environmental footprint. This would have to be taken into account in a life cycle inventory analysis that would compare production on-site with overseas manufacturing and importation of products. Environmental conditions (temperature, dust, vibration/shock e.g. from earthquakes) would be less easily managed, although the agility of the production process and the savings in producing on demand would, potentially, counteract these negative aspects.

These factors aside, in the long run it is the ability for 3DP to support communities in developing localized production strategies that, in addition to environmental sustainability, would also contribute to a triple bottom line accounting intent to maintain social sustainability. One of the significant factors emerging from this research is the identification of a need for changed relationships between those initiating the project and the local community in the development of the long-term facilities.

The transient profile of aid workers in humanitarian situations raised concerns over the optimum way that field staff might master not only the digital technology (including 3DP and 3D CAD software) to an appropriate standard, but also the collection of accurate data and testing of the prints after a print run to engineering levels prior to use. In response to this, a 'hub and spoke' model was developed where those working in situ operated in ongoing collaboration with specialists working in remote testing facilities. Importantly, such an approach was only possible because of digital communication technologies, and the ability of 3DP to support distributed manufacturing.

7 Changing Consumer Relationships

Gershenfeld, the originator of the worldwide FabLab movement, argues that the digital revolution is over, that it has been won and that the challenge going forward is to link computer science, human interaction and digital fabrication (Gershenfeld 2007). "Many of the most successful large enterprises in the world now produce goods in 'virtual global factories', with intricate spiderwebs of supply chains connecting to hundreds of other enterprises in dozens of countries" (Gore 2014, p. 4). Gershenfeld argues for digital fabrication as a means of placing advanced technologies in the hands of the wider community, with the focus on individuals and community groups rather than large enterprises, to stimulate innovation and entrepreneurship.

Thus, rather than directing users in the digital fabrication facilities he supports, Gershenfeld recommends providing the means of digital fabrication (CNC, laser cutting, 3D printing) for individuals to use as they wish. FabLabs, equipped with digital fabrication tools including 3DP (as well as electronics and sewing machines) are now available at over 150 venues around the world, including developing countries, such as Afghanistan. In conjunction with the fabrication equipment, FabLabs are also characterized by open communication links with the global FabLab network.

Hackerspaces have similar features, but they are based on Open Source principles and focus on electronics and computing rather than product design and entrepreneurship. 3D Hubs is another initiative that link individuals operating 3D printers with projects for other community members, such as through the enable project.[1] Maker communities have been growing around the world with 3DP and, as exemplified in the case study relating to the WASH project, fostering a print-on-demand, localized production mindset that disrupts the supply chain and could have significant potential benefits for contributing to sustainable production methods.

Current manufacturing and supply chain systems are based on the fundamental ideas of centralized production, the mass manufacture of common components and the subsequent distribution of finished products. The economics of mass production results in the creation of vast quantities of a design suitable to meet the needs of as large a number of customers as possible. The upfront cost of multipart molds commonly used in the predominant manufacturing process of injection molding determines the necessity to create multiple examples of the same component. Once the mold has been produced the parts can be replicated relatively cheaply. However, as Lipson and Kurman (2013, p. 25) point out, whilst there are efficiencies in the mass production method, it should be considered as a whole system: "Mass production is riddled with hidden costs and delays."

Lipson and Kurman go on to argue that, when considered in depth, factory production is laborious and unwieldy, and that the translation of a design into a mass manufacturable product inevitably impacts on the integrity of the design. They also argue that mass production means that no product is designed for a specific person, but rather for a generalized market. This reduces the ability of the manufacturer to create value added products. They also argue that, in comparison to the problems faced by manufacturers working to narrow margins, such as when producing large quantities of product just to break even, "a company's business model is based on selling small numbers of unique, constantly changing or custom-made high margin products, 3D printed production (as does the platypus) represents an evolutionary leap forward" (Lipson and Kurman 2013, p. 28).

As noted earlier, the digital revolution has provided even very small companies with the technological power that formerly would have only been characteristic of large companies which had made significant investments. Digitally-based production and distribution is making companies more agile and responsive and is allowing small companies to compete with large corporations, especially when they utilize the 'web of service' provisions in the system, such as through the metal fabrication facilities provided by Materialise in Germany. Software is also becoming more accessible, and although there is still specialized high cost software involved, this is also becoming more accessible to start-ups and small businesses.

3DP shortens the lead time of design for production: "Time to product is a key efficiency metric for companies, meaning that the shorter the time between a design and the functioning end product, the better. 3D printing shortens time to product in-hand by enabling designers and engineers to create on-the-spot product prototypes quickly and cheaply" (Lipson and Kurman 2013, p. 30). However, the emerging change in this space is more related to the development of a business

---

[1]http://enablingthefuture.org/.

case as there is a growth in iterative design production. This is where a designer uploads a new product as soon as possible, i.e. prior to the lengthy development usually required in the creation of a product that requires significant investment in molds, etc. and which takes considerable time to market.

In this new business model, customers feed back their thoughts about the product online, and the designer can then make adjustments and subsequently re-upload it for further feedback. This new relationship between the customer and the designer is a paradigm shift only made possible by the developments in digital technologies and, in particular, 3DP. It has the obvious benefit of saving cost as no investment has been made in a mold which then needs changing. The designer can, therefore, be more responsive and open to feedback and this, in turn, changes the attitudes and behaviors of the designers as well as the production practices.

A similar model is taking hold in the world of publication. Conventionally, the publication of a book could take some 2 years to reach the market from the initial acceptance by a publisher. The process also requires significant investment by the publisher in copyediting, typesetting, etc. and this makes the marketing of the book as crucial an investment as its writing and production. However, with the rise of print on demand and the emergence of ebooks, publishers such as Amazon Kindle now allow individuals to upload any form of publication without screening or delay, and collect market feedback on the viability of the product. This shift to a myriad of niche market products is echoed in 3DP.

8 Contraction and Convergence

There is currently an unfair distribution of wealth across the planet. From this starting point alone, the need for contraction and convergence is paramount. This is not only from a moral standpoint that reflect ideas of universal fairness, but also from the perspective of reducing conflict and the negative impact that this has on societies, infrastructure and the environment. As the USA loses its position as the arbiter of world peace, there is an increasing need for Western societies to encourage economic stability in countries around the world to avoid further conflict. With economic stability and improved wealth, countries are better positioned to support the welfare of their citizens and there is less need for economic migration and the burden that this places on the infrastructure of cities from border countries.

Environmentally, the perpetrators of the majority of the damage (i.e. companies) are predominantly from the developed countries. Meanwhile, emerging economies are emulating the practices of developed countries with, possibly, even fewer constraints. As the world population expands exponentially, the need to rethink radically every aspect of living on the planet for the sake of sustainability—and survival—is increasingly urgent.

The Industrial Revolution emerged in response to perceived needs in the eighteenth century, and was informed by the limited knowledge and understandings of the period. It was also influenced by the social structures of the time. The current social structures that are supported by the economic patterns in Western societies that have developed since the Industrial Revolution are being challenged for the first time since the Industrial Revolution by the digital revolution and the opportunities it provides. The development of digital technologies such as those in communications, data generation, computer modelling and digital fabrication are now converging, and this integration is allowing for a rethink of humans' interaction with the physical environment, and the patterns of production and consumption within it. In so many ways, Western-style manufacturing practices are about as bad for the environment as it is possible to be. Indeed, if the aim was to create a system that would waste natural resources, irreversibly pollute nature, over-invest on the embodied energy in a product, and create excessive waste, then current economic production methods and systems have surely succeeded.

Whilst the greed of individual entrepreneurs, and even the well-meaning attempts at social engineering sponsored by manufacturing such as those evident in the history of Unilever and Port Sunlight in the UK, could have, in the past, been viewed as ignorant of the wholescale damage their operations were causing, few could be unaware of the impact of production and consumption today. Yet the attitude towards waste and the negligence over environmental concerns such as the production of methane in landfill from waste food and the distressing consequences of the abundance of plastic bottles in the waste stream on marine life is shockingly immoral. To create such damage for the sake of having fully stocked supermarket shelves (Gore 2014) right up until closing time, or merely to have 700 mL of water to drink with maximum convenience, is a sad indictment of societies' attitudes. Furthermore, hoping that companies voluntarily follow the lead of pioneers such as Ray Anderson at Interface[2] carpets or of the Agency of Design,[3] in shifting to Product Service Systems thinking (Hawken et al. 2013) is fairly futile.

If this is the case, then a step-change is required. The likelihood is that this can only come about when the world is at the brink of destruction—if at all. However, if there is to be a change, then the digital revolution is providing the tools with the potential to make it happen. Digital communication is externalising repositories of collective wisdom in the same way that writing first did, but in more powerful manifestations. This allows for the emergence of the Global Mind that Gore refers to. This is not only about pooling knowledge, but also about setting up digital communication systems for decision-making that are external to the individual. This could be a force for good, although it also has the potential to create as many problems as it could solve. However, assuming the best, then the externalising of decision-making based on Big Data fed from worldwide sources could potentially mean that the impact of behaviour on one side of the world as they result in change on the other could finally be made visible.

---

[2]https://www.interface.com/US/en-US/global.

[3]http://www.agencyofdesign.co.uk/.

By creating these visible links, the morality of the privileged individual could potentially be redirected. In the same way that Web 2.0 sites such as Wikipedia are policed by the majority in a way that would not have seemed possible or likely 10 years ago, so too could the real world impact of decisions made by individuals casually throwing their plastic water bottles away in a Western country be seen in relation to the impact it has elsewhere. For example, RFID tags and GPS could feasibly mean that a plastic bottle cap from a water bottle drunk in the UK and irresponsibly dumped could be tracked into the ocean and to the stomach of a young albatross. Legislation is spreading that creates end-of-life responsibility for manufacturers of goods such as cars and fridges. Digital technology could take this further by creating a sensor-based tracking approach for the interaction of people with every physical product they come across. There are already new apps that have been developed which track drivers habits to give them a 'safe driver' and 'eco-driver' rating system (such as Green Driver,[4] Skytrack[5] and Eyetrack[6]), and these can be shared online for crowd policing and behaviour reinforcement. The resulting collective conscience would then become part of the Gore's Global Mind and new sustainability sensitivities not previously evident could become possible.

9 New Patterns of Production

Assuming a global collective conscience with sustainability sensitivities based on the linking of individual actions to the complexities of consequences worldwide, then a complete redesign of the systems supporting human interactions with their physical environment could occur. If this happened—either stimulated by a visible crisis or by the visualization of global data creating a worldwide reaction—then 3DP would be a significant part. In fact, it is only the possibilities that 3DP are now providing that allow for speculation over how the changing production consumption relationships could be redesigned: "3D printing is the catalyst that cloud manufacturing has been waiting for. Cloud manufacturing will be a decentralized system, built on a foundation of ultra-large networks of small manufacturing companies" (Lipson and Kurman 2013, p. 46).

Imagine a future where supply chains are based on sustainability aims and enabled by digital technologies. Ryan (2004) argues that society needs to address sustainability based on six strategic principles: (1) valuing prevention, (2) preserving and restoring 'natural capital', (3) life cycle thinking (closing system cycles), (4) increasing 'eco-efficiency' by 'factor x', (5) decarbonising and dematerialising

---

[4]http://www.donlen.com/uploadedFiles/Home/The_Donlen_Difference/Green_Solutions/WaterOne-GreenDriver-GreenFleetMag.pdf.

[5]http://www.skytrack.net/index.php/en/applications/ekodrive.html.

[6]http://eyetrak.co.uk/expert-spotlight-on-european-eco-driving/.

the economy, and (6) focusing on design—of products and product-service. Considering these principles in the context of a digital revolution, rather than based on those underpinning the Industrial Revolution, creates new models of production and consumption that turn current systems on their head. The most significant paradigm shift that 3DP has the potential to support (and which responds to Ryan's propositions) is in relation to valuing prevention and preserving natural capital.

Valuing prevention as a strategy is based on preventing any environmental impacts as early in the supply chain as possible. 3DP shifts production from mass production to an on-demand approach that clearly supports a valuing prevention strategy, which is based on the preservation of natural capital—reducing the material used—and reducing the impact of production processes (by-products, etc.) through reduced quantities of product being made. In addition, 3DP supports the development of 'lightweighting' for product design and for the optimization of product functions through building the object around the optimal forms, rather than cutting channels into outside forms (e.g. for cooling design elements in industrial objects). This reduces the material used in each object and therefore contributes to minimising the sourcing of raw materials, their processing and end-of-life impact.

Stewart Walker (2006), in *Sustainability by Design*, suggested that a sustainability strategy would be to improve the connection between people and products. He observed that, by becoming invested in the products, people would develop an ongoing relationship with them which would counteract the attitude of obsolescence that dominates consumer culture. One of the significant factors in the use of 3DP is that, because it does not rely on an initial mould, as is the case with many other manufacturing processes used in mass production, each product can be unique. Combined with this is the fact that the print is created from a 3D computer model. Therefore it can be modified to meet customization requirements driven by data sets. This customization of the product can potentially contribute to creating an invested product. This would then potentially be kept longer by the customer (who has invested in it), and even repaired as needed, extending its useful life and the length of time its embodied energy would need to be assessed against as part of its life cycle assessment.

A further aspect is that customized products are likely to be more effective as their function could be specifically tailored to the task and the user. Therefore they are, once again, likely to be kept rather than replaced, thereby reducing the number of products produced for a user overall. Finally in this context, the relationship of producers and consumers facilitated by digital communication and materialized by 3DP (as found on retail websites such as Nervous System, see Footnote 1), would increase the emotional investment made by the user, and the likelihood of optimum function and usability for the user. Once again, this would increase the retention of the product and postpone its disposal.

If it was possible for the wholesale redesign of production systems at this point, based on the information societies now have, and informed by the technological developments of the last 30 years, then it could be theoretically possible to create patterns of manufacturing and consumption that operated within sustainable levels—at least until over-population caused the implosion of the species. In this world, no more raw materials based on non-renewable resources would be sourced unless there were exceptional circumstances. Instead, all the metals and thermoplastics used so far would be mined from landfill. Designs would be created in collaboration with individuals, and informed by expertise from around the world connected by the Internet.

Each design would be value-added—invested through technology and time and skill to meet the specific needs of the user, based on data generated through multiple technologies, such as scanning. Each design would be lightweight, based on complex software applications, such as the current 3-matic by Materialise, to optimize the use of materials, and designed for function with no additional wasted material. The designs would be created wherever the expertise was based, and the digital file (STL) transmitted directly to the point of use, or to the nearest point where the machinery to print the object was housed, wherever in the world that was in a distributed manufacturing system. This would save the significant environmental impact of transport miles in relation not only to the product itself but also to the packaging. Support material would be reused—albeit down-cycled—as much as possible, and products would be specifically designed for 3DP to reduce scaffolding and nested to minimize the use of virgin powder in selective laser sintering builds.

To support this sustainability ideal further, products would be designed to be repairable. Desktop printers would become ubiquitous in peoples' homes, and this would foster the rise of a 'Maker society' (Anderson 2014) well equipped to interact with products and to conduct minor repairs. Products would also be designed for disassembly so that where multiple materials were used the product would be designed to enable materials to be separated out. Where possible, products would be created from a single, recyclable material, and designs created so that they could be taken apart and revitalized for reuse.

At the end of life, the product would be disassembled and the materials reclaimed and recycled. There would inevitably be some downcycling, but there would be systems in place to downcycle the material from food safe applications down to products, furniture and finally construction products (Thompson 2013). This would need to be investigated for unforseen concerns but, as discussed in cradle to cradle (Braunguart and McDonough 2002), the products would be designed with this understanding in mind to mitigate any downstream issues.

There would inevitably be hidden costs to this approach, not least the dismantling of current systems of production that go far beyond the structure of buildings and equipment, but also the social fabric that has been created around conventional manufacturing facilities. The impact on economies in the short term could be catastrophic. However, in the long term, the shift from mass production controlled by the minority of wealthy people on the planet living in developed countries, and the environmental and social cost of the majority in developing and emerging economies, could result in saving the ultimate cost of the loss of the human habitat—not only through reducing environmental impact, but also through supporting contraction and convergence in an increasingly linked, borderless world.

10 Conclusion

The sustainability movement began for the industrial designer as eco-design. However, the world is moving too rapidly towards crisis point for compromises of eco-design. It is only the hope of the digital technologies allowing for a new global mind collective operating in concert that provides even the potential, and indeed hope, for avoiding what seems inevitable. However, history also shows that huge behaviour changes are possible in response to changing imperatives and technologies—such as can be seen in the urbanization of populations following the Industrial Revolution.

> In our own lives, we are accustomed to gradual, linear change. But sometimes the potential for change builds up without being visibly manifested until the inchoate pressure for change reaches a critical mass powerful enough to break through whatever systemic barriers have held the change back. Then suddenly one pattern gives way to another that is entirely new. This 'emergence' of systemic change is often difficult to predict, but does occur frequently both in nature and in complex systems designed by human beings (Gore 2014, p. 29).

The dematerialising of the economy is a possibility if the digital really proves to be the next evolutionary stage of the human species. There will still be a need for objects—even within a dematerialized society. However, the shift is towards co-design, bespoke production that maximizes resources and extends the output from embodied energy with design for disassembly, flexibility and repair. Transport miles should become a thing of the past with distributed manufacturing and the supply chain being replaced by a value chain based on product service system thinking.

If there were a chance to reduce the industrial revolution to, as Aldersey-Williams (2011) puts it, a 'blip' in a world of cottage production, then 3DP would be an integral part of that scenario. Already 3DP is responsible for changing business models. The rise of online service providers, such as Shapeways with hundreds upon thousands of prints for sale by individuals, and evidence of completely new thinking about business practice, such as that of Hasbro working with customers to allow for additional licensing as a business strategy, rather than protecting their IP as previously, suggest that 3DP is part of the changes that are happening through the reorganization of business using the Internet, and the disruption of conventional supply chain and logistical practices.

According to Kozel (2013, p. 7), the main focus in considering groundbreaking design change is: "goods that reflect not only cultural and technical phenomena but also social developments". The products being created using 3DP are groundbreaking because they disrupt conventional manufacturing in ways that have the potential to impact on each stage of the supply chain. For the concept of environmental sustainability to be met as outlined by the Bruntland commission, a radical redesign of entire systems of production, distribution, use and disposal needs to be created. The sustainability potential of a relatively dematerialized society can, however, be made possible by the triumph of product service systems and on demand 3DP over mass production. This, in turn, provides an opportunity to bring humanity back from the brink of extinction brought about by its pollution of the environment. Thus, 3DP provides an opportunity for step-change that could contribute to the survival of humanity.

References

Aldersey-Williams H (2011) The new tin ear: manufacturing, materials and the rise of the user-maker. RSA, London

Anderson C (2014) Makers: the new industrial revolution. Crown Business, New York

Christopher M (2011) Logistics and supply chain management, 4th edn. Pearson Education Ltd., Harlow

Christopher M, Holweg M (2011) Supply chain 2.0: managing supply chains in the era of turbulence. Int J Phys Distrib Logist Manag 41(1):63–82

Doheny M, Nagali V, Weig F (2012, May 1–7) Agile operations for volatile times. McKinsey Quarterly

Durgavich J (2009) Customs clearance issues related to the import of goods for public health programs. US AID. Available via http://deliver.jsi.com/dlvr_content/resources/allpubs/policypapers/CustClearIssu.pdf. Accessed 25 Sept 2015

Economist (2013) Advanced manufacturing—adding and taking away. The Economist, 31 Dec 2013. Available via http://www.economist.com/blogs/babbage/2013/12/advanced-manufacturing. Accessed 25 Sept 2015

Elkington J (1999) Cannibals with forks: Triple bottom line of 21st century business. Capstone, Minnesota

Fawcett S, Waller M (2014) Supply chain game changers—mega, nano, and virtual trends—and forces that impede supply chain design (i.e., building a winning team). J Bus Logist 35(3):157–164

Flannery T (2007) The weather makers: Our changing climate and what it means for life on earth. Penguin, London

Forty A (1992) Objects of desire: design and society, 1750–1980. Revised ed. Thames and Hudson, London

Fuad-Luke A (2005) The eco-design handbook: A complete sourcebook for the home and office. 2nd revised ed. Thames & Hudson, London

Gartner (2014) Gartner says worldwide shipments of 3D printers to reach more than 217,000 in 2015. Gartner Inc. Available via http://www.gartner.com/newsroom/id/2887417. Accessed 25 Sept 2015

Gershenfeld N (2007) Fab: the coming revolution on your doorstep—from personal computers to personal fabrication. Basic Books, New York$$

Gore A (2014) The future. W.H.Allen, London

Hawken P, Lovins A, Lovins L (2005) Natural capitalism: creating the next industrial revolution, 2nd edn. US Green Building Council, Washington

Kozel N (2013) Design: the groundbreaking moments. Prestel, New York

Lipson H, Kurman M (2013) Fabricated: the new world of 3D printing. Wiley, New Jersey

Loy J, Tatham P, Healey R, Tapper C (2015) 3D printing meets humanitarian design research. In Handbook of Research on Creative Technologies for Multidisciplinary Applications. IGI Global, Hershey

Lusch R, Vargo S (2014) Service-dominant logic: premises, perspectives, possibilities. Cambridge University Press, Cambridge

Mau B, Institute Without Borders (2004) Massive change: A manifesto for the future of global design. Phaidon Press, London

McDonough W, Braungart M (2002) Cradle to cradle. North Point Press, New York

Moore   C   (2009).   https://www.ted.com/talks/capt_charles_moore_on_the_seas_of_plastic. Accessed 20 Sept 2015

Packard V (2011) The waste makers. Reprint Ed. IG Publishing, Singapore

Pannett L (2014, Jan) 3D: the future of printing. Supply Manag 34–37

Papanek V (2005) Design for the real world: human ecology and social change, 2nd edn. Chicago Review Press, Chicago

Ryan C (2004) Digital eco-sense: sustainability and ICT—a new terrain for innovation, Lab 3000, Melbourne

Shell (2008) Shell energy scenarios to 2050. Available via http://s01.static-shell.com/content/dam/shell/static/future-energy/downloads/shell-scenarios/shell-scenarios-2050signalssignposts.pdf. Accessed 22 Sept 2015

Shinall J (2013) Companies large and small are using 3-D prototyping to push the boundaries of innovation. USA Today. Available via http://www.usatoday.com/story/tech/2013/03/20/3d-printing-apple-samsung-jabil-ford-maker-autodesk/1973753/. Accessed 25 Sept 2015

Singh M (2009) In times of uncertainty, focus on the future. Supply Chain Manag Rev 1(3):20–26

Tatham P, Loy J, Peretti U (2015) Three dimensional printing—a key tool for the humanitarian logistician. J Humanit Logist Supply Chain Manag 5(2):188–208

Thompson R (2013) Sustainable materials, processes and production. Thames & Hudson, London

Vargo S, Lusch R (2004) Evolving a new dominant logic for marketing. J Market 68(January):1–17

Walker S (2006) Sustainability by design: Explorations in theory and practice. Routledge, London

Printed in the United States
By Bookmasters